Environmental Biology

Routledge Introductions to Environment Series

Published and Forthcoming Titles

Titles under Series Editors:
Rita Gardner and Antoinette Mannion

Environmental Science texts

Environmental Biology
Environmental Chemistry and Physics
Environmental Geology
Environmental Engineering
Environmental Archaeology
Atmospheric Systems
Hydrological Systems
Oceanic Systems
Coastal Systems
Fluvial Systems
Soil Systems
Glacial Systems
Ecosystems
Landscape Systems

Titles under Series Editors:
David Pepper and Phil O'Keefe

Environment and Society texts

Environment and Economics
Environment and Politics
Environment and Law
Environment and Philosophy
Environment and Planning
Environment and Social Theory
Environment and Political Theory
Business and Environment

Key Environmental Topics texts

Biodiversity and Conservation
Environmental Hazards
Natural Environmental Change
Environmental Monitoring
Climatic Change
Land Use and Abuse
Water Resources
Pollution
Waste and the Environment
Energy Resources
Agriculture
Wetland Environments
Energy, Society and Environment
Environmental Sustainability
Gender and Environment
Environment and Society
Tourism and Environment
Environmental Management
Environmental Values
Representations of the Environment
Environment and Health
Environmental Movements
History of Environmental Ideas
Environment and Technology
Environment and the City
Case Studies for Environmental Studies

Routledge Introductions to Environment

Environmental Biology

Allan M. Jones

London and New York

First published 1997
by Routledge
11 New Fetter Lane, London EC4P 4EE

Simultaneously published in the USA and Canada
by Routledge
29 West 35th Street, New York, NY 10001

Transferred to Digital Printing 2004

Typeset in Times by Keystroke, Jacaranda Lodge, Wolverhampton

British Library Cataloguing in Publication Data
A catalogue record for this book is available from the British Library

Library of Congress Cataloging in Publication Data
Jones, A. M. (Allan M.)
Environmental biology / Allan M. Jones.
p. cm. — (Routledge introductions to environment)
Includes index.
1. Ecology. I. Title. II. Series.
QH541.J655 1997
577—dc21 96–47087
CIP
ISBN 0–415–13620–2
0–415–13621–0 (pbk)

This book is dedicated to my family, Angela, David and Gordon, in the hope that it will encourage them in all they do, and to my parents, without whose sacrifices I could never have written this book.

Contents

Series editors' preface
Environmental Science titles

The last few years have witnessed tremendous changes in the syllabi of environmentally-related courses at Advanced Level and in tertiary education. Moreover, there have been major alterations in the way degree and diploma courses are organised in colleges and universities. Syllabus changes reflect the increasing interest in environmental issues, their significance in a political context and their increasing relevance in everyday life. Consequently, the 'environment' has become a focus not only in courses traditionally concerned with geography, environmental science and ecology but also in agriculture, economics, politics, law, sociology, chemistry, physics, biology and philosophy. Simultaneously, changes in course organisation have occurred in order to facilitate both generalisation and specialisation; increasing flexibility within and between institutions is encouraging diversification and especially the facilitation of teaching via modularisation. The latter involves the compartmentalisation of information which is presented in short, concentrated courses that, on the one hand are self contained but which, on the other hand, are related to pre-requisite parallel, and/or advanced modules.

These innovations in curricula and their organisation have caused teachers, academics and publishers to reappraise the style and content of published works. Whilst many traditionally-styled texts dealing with a well-defined discipline, e.g. physical geography or ecology, remain apposite there is a mounting demand for short, concise and specifically-focused texts suitable for modular degree/diploma courses. In order to accommodate these needs Routledge have devised the Environment Series which comprises Environmental Science and Environmental Studies. The former broadly encompasses subject matter which pertains to the nature and operation of the environment and the latter concerns the human dimension as a dominant force within, and a recipient of, environmental processes and change. Although this distinction is made, it is purely arbitrary and is made for practical rather than theoretical purposes; it does not deny the holistic nature of the environment and its all-pervading significance. Indeed, every effort has been made by authors to refer to such interrelationships and to provide information to expedite further study.

This series is intended to fire the enthusiasm of students and their teachers/lecturers. Each text is well illustrated and numerous case studies are provided to underpin general theory. Further reading is also furnished to assist those who wish to reinforce and extend their studies. The authors, editors and publishers have made every effort to provide a series of exciting and innovative texts that will not only offer invaluable learning resources and supply a teaching manual but also act as a source of inspiration.

A. M. Mannion and Rita Gardner
1997

Series International Advisory Board

Note on the text

Bold is used in the text to denote words defined in the Glossary. It is also used to denote key terms.

Figures

Tables

Boxes

Acknowledgements

I would like to thank my colleagues for their valuable comments at various stages of the preparation of this book. In particular, I would like to thank David Hopkins and Cathy Caudwell for commenting on an early version of the book and Rachel Morris for assistance in preparing the glossary and the final version of the text. I would also like to thank my sons, David and Gordon, for their assistance in the preparation of many of the figures; their imagination and lack of inhibition when using CorelDraw were impressive. Most figures were drawn using CorelDraw 5, a product of Corel Corporation, whose use is gratefully acknowledged. The use of clipart components provided by Totem Graphics Inc. and Image Club Graphics Inc. is also gratefully acknowledged. My wife, Angela, also helped enormously both in her patience and in reading various parts of the text. Finally, I would like to thank the editors and referees for their valuable comments at various stages of production.

An introduction to environmental biology

Any understanding of how our environment works and reacts requires a knowledge of living organisms and the systems in which they participate. This book introduces the core elements of biology that students studying the broader topic of environmental science need to appreciate for an understanding of the role of organisms in the environment. Although intended for the non-specialist, the text requires the appropriate terminology to describe the various processes that underlie this understanding. However, I have kept technical terms to a minimum and provided definitions in the glossary. The terminology used is a compromise intended to be sufficient as a starting point for those wishing to acquire a deeper understanding of any particular topic.

It is impossible to overemphasise the importance of the biosphere and its diverse activities. Indeed, our oxygen-rich atmosphere is a consequence of the evolution of photosynthesis in the distant past: until that oxygen-producing process evolved, the Earth's atmosphere was very different and incapable of supporting larger organisms. James Lovelock expressed this role in his philosophy of Gaia, the concept of the Earth as a complex living organism composed of the component parts that we recognise as biosphere elements. The practical applications of a knowledge of environmental biology lie in the fields of environmental history, geography, land use and development, environmental impact assessment, agriculture and forestry, climate change, pollution and resource conservation. This book emphasises the key biological processes necessary for a better understanding of the role of organisms in these fields of study.

The text attempts to build logically from the fundamentals of living organisms and their organisation and distribution through increasingly complex components, namely individuals, populations, communities and ecosystems. I have tried always to introduce topics by providing the background needed for their understanding. I have tried also to balance the aquatic and terrestrial exemplars, aquatic systems often being neglected in biogeographical texts. The content of the chapters presented is limited to the more general aspects of most topics because the key topics are the subject of specialist volumes in the Environmental Science and Environmental Studies series. This results in a compromise between superficiality and depth of treatment and it is important that other volumes in this series are seen as extensions of the key topics discussed here.

The basic organisation of living forms

Key concepts

- **Living things combine many vital characteristics to distinguish them from non-living materials.**
- **Cells are the fundamental units of life which interact with their environment.**
- **The cell contains various structures which organise its activities.**
- **Energy transformations form the engine of the cell.**
- **The materials of life: eleven main elements, water and the macromolecules.**
- **The diversity of life and its levels of organisation.**
- **The species concept and the kingdoms of life.**

There are a number of features and concepts which apply to the study of all living things. The construction and activities of cells underlie any understanding of the material and energetic requirements of life at all levels of study. The development and evolution of the interrelationships between the cell and its environment have resulted in increases in complexity and diversity. This chapter briefly reviews the key organisational features of living forms as related to environmental systems and outlines the ways in which biologists define and describe the variety of living forms.

1.1 The fundamentals of life

Some of the basic properties of living material need to be considered if we are to understand the ways in which organisms interact with their environment. Organisms are essentially chemical machines based around the chemistry of carbon compounds and of water, possibly the most important molecule to life.

1.1.1 The characteristics of life

Life is not a simple concept and it is impossible to define precisely what life is: all that can be done is to describe the observable phenomena that distinguish living matter from inanimate matter. Living organisms are characterised chemically by their ability to interact with their environment. Both biotic (living) and abiotic (non-living) components exchange and synthesise materials and energy in carefully controlled manipulations through the integrated chemical processes called metabolism.

Most properties exhibited by living forms are found also in some form in non-living material, i.e. they are characteristic of living material but not restricted to it. These include:

- **Cellular organisation**: all organisms comprise one or more cells but non-living structures such as coacervates, a mass of minute (colloidal) particles held together by electrostatic attraction, also appear cell-like in organisation.
- **Nutrition**: the acquisition of materials and energy necessary for growth. In living material this is either
  - **autotrophic**, using light energy (**photosynthesis**) or chemical energy (**chemosynthesis**), or
  - **heterotrophic**, obtaining materials and energy by the breaking down of other biological material using digestive **enzymes** and then assimilating the usable by-products.

  Non-living forms, e.g. stalactites, acquire material only passively.
- **Growth**: non-living objects such as crystals usually grow by the addition of materials to their exterior while living material grows from within using materials acquired through nutrition.
- **Respiration**: a series of energy-producing chemical reactions releasing energy from energy-rich compounds either by 'burning' with oxygen (aerobic respiration) or by reactions not involving oxygen (anaerobic respiration and fermentation).
- **Responsiveness (irritability)**: the ability to respond to changes in both internal and external environments, usually to improve the chances of survival. This may involve growth (plants), movement (mobile species), or physiological changes. Inanimate material can respond only passively and to external forces.
- **Movement**: non-living material moves only as a result of external forces while living material moves as a result of internal processes at cellular level or at organism level (locomotion in animals and growth in plants).
- **Excretion**: the elimination from living organisms of waste by-products from metabolism.
- **Reproduction**: all living forms can multiply, thus ensuring the continuation of the species. Non-living material can sometimes replicate, e.g. the production of crystals and the spreading of a fire. However, it never involves the use of hereditable material (DNA or RNA; see Section 1.1.3.4), which contains a complex coding for the construction, development and functioning of the progeny.

These observable characteristics are typical of all living organisms but the real significance of living material lies in its ability to extract, convert and use energy and materials to maintain and even increase its energy content. The process of metabolism, a characteristic of life, is itself divided into two processes:

- **Catabolism**, the breaking down of complex materials into simpler ones using enzymes and releasing energy, and
- **Anabolism**, the utilisation of energy and materials to build and maintain complex structures from simple components.

Non-living material can only undergo catabolic activity according to the Second Law of Thermodynamics (Box 1.1).

Box 1.1

The flame of life – an interesting analogy

Non-scientific literature often uses the phrase 'The flame of life'. However, this analogy is not as strange as it may seem at first sight since the properties of a burning flame parallel many of the characteristics of living things. Thus, arguably, a flame:

- *respires*: it requires oxygen for its chemical conversion of organic material into carbon dioxide, water and energy (heat). This is fundamentally the same catabolic process carried out in all aerobic cellular respiration. However, cells carry it out under controlled conditions using enzymes and with the controlled release of energy into usable chemical forms. Ultimately, the energy becomes dissipated as heat. Remove the oxygen from an aerobic cell or a flame and either will die.
- requires *nutrition*: a flame requires a fuel source just as a cell does. The flame consumes flammable materials which are the equivalent of the food of a heterotroph.
- *reproduces*: provide a flame with extra resources (e.g. more wood) and it will spread in a manner equivalent to asexual reproduction by fission.
- *excretes*: when a flame burns, it excretes carbon dioxide, water and ash.
- *grows*: with increased resources, flames will grow larger.
- *moves*: flames are never still and move extensively although this movement is the result of external forces only.
- is *irritable*: flames are sensitive to external forces and will respond to changes in the external environment, e.g. a candle will respond to the pressure changes in sound waves.
- is *organised*: a flame has a structure related to regions of different temperature, etc. Look carefully at a candle flame and you will see it appears organised!

Clearly a flame is not alive but how can this be proved? The most decisive criteria to apply are:

1. Non-living materials never replicate using hereditable materials (DNA or RNA).
2. Non-living material cannot carry out **anabolic** metabolism, i.e. make complex materials from simpler ones. This is the basis of biosynthetic processes carried out during repair and growth in all living organisms. Catabolic processes occur in both living and non-living materials.

Living organisms are best defined as structures exhibiting anabolic metabolism and reproduction using hereditable materials. However, their diversity derives from the ways in which the characteristics of life described above have been exploited by different processes and mechanisms developed at the various levels of organisation and in different environments. This has resulted in enormous structural and functional diversity in living forms, the details of which are part of other specialised disciplines such as Zoology, Botany and Microbiology. This text deals mainly with the functional aspects of the dynamic interactions between organisms and their environment. However, the significance of structural features cannot be ignored if the functional relationships are to be fully understood. There is an intimate relationship between diversity of form and function.

1.1.2 The cell: the fundamental unit of life

The diverse array of materials and processes that make up a living organism are structured and organised to provide a co-ordinated system. The basic unit that does this is the cell, an elaborately organised chemical machine that is the smallest unit of life. Today, the study of cell biology forms a discipline of its own and this book summarises only the salient features.

Cells may exist as independent units, e.g. the Protozoa, or as parts of multicellular organisms in which the cells may develop specialisations and form tissues and organs with specific purposes. Cells are of two fundamental types, prokaryotic and eukaryotic (see also Section 1.2.4). **Prokaryotic** (pro = before; karyotic = nucleus) cells are simpler in design than eukaryotic cells, possessing neither a nucleus nor the organelles found in the cytoplasm of eukaryotic (eu = true) cells (Table 1.1 and Figure 1.1). Organelles are internal cell structures, each of which has a specific function within the cell.

Table 1.1 ***Comparison of key features of prokaryotic and eukaryotic cells***

	Prokaryotic	*Eukaryotic (plant)*	*Eukaryotic (animal)*
Size	1–10 μm diameter	10–100 μm diameter	10–100 μm diameter
Plasma membrane	+	+	+
Cell wall	usually (peptidoglycan)	+ (cellulose)	–
Nuclear envelope	–	+	+
Nucleolus	–	+	+
DNA	+ single loop	+ (chromosomes)	+ (chromosomes)
Mitochondria	–	+	+
Chloroplasts	–	+	–
Endoplasmic reticulum	–	+	+
Ribosomes	+	+	+
Vacuoles	–	+ (usually large, single)	+ small
Golgi apparatus	–	+	+
Lysosomes	–	often	usually
Cytoskeleton	–	+	+
9+2 cilia or flagella	–	not in angiosperms	often

Note: + = present; – = absent.

Cells are small structures, prokaryotic cells being between 1 and 10μm in diameter while eukaryotic cells may be much larger, typically from 10 to 100μm in diameter. The limitations on size are related to the surface:volume ratio. The rate of exchange of materials into and out of metabolising cells depends on the relationship between the surface area of the plasma membrane surrounding the cell and the volume of the cell cytoplasm. A cell needs a surface area sufficient to exchange materials with the environment and thus a high surface to volume ratio. Size limits are also related to the fact that movement of materials within the cytoplasm is by diffusion, which becomes less effective as cell volume increases. However, a few cells do attain larger sizes either

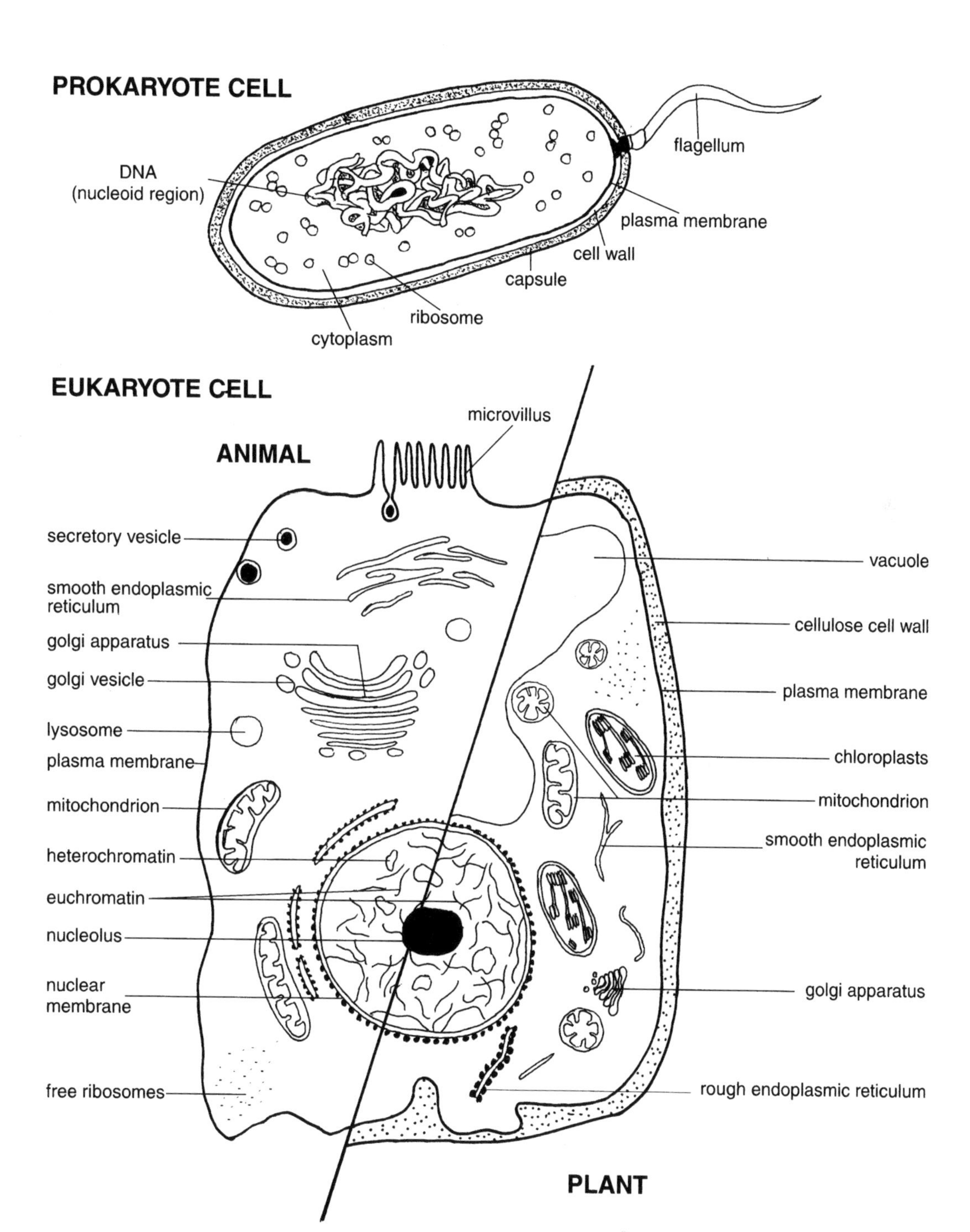

Figure 1.1 ***A diagrammatic comparison of prokaryotic and eukaryotic (plant and animal) cell structures, showing the main subcellular components only***

- by modifying the cell shape to increase the surface to volume ratio, e.g. a nerve cell becomes long and thin, or
- by having convoluted cell margins as in many protistans.

1.1.2.1 The plasma membrane: interface with the environment

The plasma membrane is vital to the functioning of cells, separating the cell from its non-living environment. It is bounded externally in plants by a more rigid structure, the plant cell wall (Box 1.2). The plasma membrane is a lipid bilayer with a complex assembly of many proteins loosely attached to the membrane. Its role is to control the cell's interactions with the environment, including:

- movement of water to which the membranes are freely permeable
- incorporation of bulk items (particulate food and liquids) into the cell
- selective transport of molecules and ions
- reception of chemical messages (e.g. hormones)
- attachments to other cells.

Any environmental factors that disrupt this membrane will cause cell death. All except animal cells have supportive cell walls external to the plasma membrane (Box 1.2).

The movement of material across membranes occurs in three ways.

- Passive transport results from concentration differences across the membrane and the consequent concentration gradient. There are three forms: diffusion, osmosis and facilitated diffusion. *Diffusion* is the random movement of molecules from an area of higher concentration to one of lower concentration. *Osmosis* is the diffusion of water across a selectively permeable membrane that permits free passage of water but not that of one or more of the solutes. Facilitated diffusion is the transport of specific molecules across a membrane by a carrier protein, a mechanism often called a channel.
- Active transport mechanisms require energy because they are usually operating against a concentration gradient or require membrane modifications. Such mechanisms involve metabolically operated pumps, vital processes because they allow cells to concentrate and collect molecules against concentration gradients using chemical energy. Three types of pump are known: the sodium–potassium pump, the proton pump and coupled channels.
- Endocytosis is the process whereby cells engulf minute particles of food or fluid by enfolding them within vesicles, minute pinched-off enclosures formed from the plasma membrane. It occurs as *phagocytosis*, the incorporation of particles such as bacteria by mammalian white blood cells and *pinocytosis*, the incorporation of extracellular fluids.

1.1.2.2 Cell organelles and their functions

Table 1.2 presents a summary of the main organelles (sub-cellular structures) found in eukaryotic cells and their functions. Cells are organised internally to facilitate the

Box 1.2

Plant cell walls

The primary plant cell wall contains large amounts of the polysaccharide *cellulose*, formed into microfibrils by lateral chemical bonding. These microfibrils are effectively cemented together by other polysaccharides called hemicelluloses. A third class of substances, pectins, is used to bind together adjacent cell walls. Where vascular plants have cell walls that need to be particularly strong, a secondary cell wall is added. This is much thicker than the primary wall and is usually impregnated with **lignin**, making it very strong. Lignins are chemically complex and resist chemical, fungal and bacterial attack very effectively. Only white rot fungi (basidiomycetes) are capable of complete mineralisation of lignin and, therefore, wood. The difficulty of degrading these materials is shown by the observation that some fractions of soil organic matter have been radiocarbon dated to over a thousand years old. Several other features of plant material diminish its degradability, notably the development of waxy cuticles over the leaves and incorporation of tannins into the cells of conifers.

Plant cell walls are ecologically important structures as well as being an important part of the structural integrity of the plants themselves. This is because much dead plant material is resistant to the decomposition processes which release and recycle nutrients and their role in the structuring of soils and sediments. Plant materials are usually the most important input of organic matter to soils. It has been estimated that up to 70 per cent of plant material is never incorporated into the herbivore/carnivore food chain but goes directly to the detrital food web.

Dead plant material is generally referred to as litter and its breakdown is due to both physical and biological factors. The table below shows the quantities of accumulated dead organic matter in different forest types showing that it is greatest at high latitudes and tends to diminish towards lower latitudes. This is related both to the dominance of the conifers with their tough needle-like leaves at these higher latitudes and to the physical conditions that prevail. Conifer forests produce very acid conditions in the soil which act to diminish the abundance of the soil flora and fauna that develop under normal conditions of pH. Thus such materials tend to accumulate, especially since low temperatures also inhibit decomposition in many such areas.

Table B1.1 ***Relative accumulation of plant litter in selected ecosystems***

Forest type	*Mass of organic matter (tons/hectare) in forest layer*
Shrub tundra	85
Taiga	30–45
Broad-leaved forest	15–30
Subtropical forest	10
Tropical rainforest	2

The toughness of plant cell walls has also required the development of chemical and structural adaptations for grazing and digestion in the herbivore and detritivore components of any system. Box 4.8 deals with this aspect.

Table 1.2 Summary of the main functions of cell organelles

Organelle	*Main function/s*
Nucleus	Control of cell activity; storage of DNA
Endoplasmic reticulum	Protein synthesis; distribution of material
Ribosomes	Sites of protein synthesis
Mitochondria	Chemical energy conversions for cell metabolism
Plastids (plants cells only)	Conversion of light energy into chemical energy; storage of food and pigments
Golgi complex	Synthesis, packaging and distribution of materials
Lysosomes	Digestion, waste removal and discharge
Vacuoles	Digestion and storage
Microfilaments and microtubules	Cell structure; movement of internal components
Cilia and flagella	Locomotion and production of external currents
Microbodies	Chemical conversions; discharge

various biochemical processes taking place side by side. They achieve this by the extensive use of membranes (**endoplasmic reticulum**) which serve to localise and integrate these activities. Many organelles are discrete structures with a surrounding membrane whilst others form a system of folded membranes that provide extensive surfaces for the organisation of chemical reactions. They also partition the cell into smaller compartments to minimise metabolic interference.

Plastids are a group of organelles found only in plant cells and are composed of two outer membranes enclosing a fluid, the stroma. The most important plastid is the chloroplast, the organelle that carries out photosynthesis, converting light energy into chemical energy. Both plastids and mitochondria contain their own DNA, which is itself similar to prokaryote DNA. This observation has given rise to a number of hypotheses on the origin of the eukaryote organelles. The widely supported Endosymbiosis theory proposes that such organelles were once free-living prokaryotic cells that took up residence inside a larger prokaryotic cell.

1.1.2.3 Energy transformations – the engine of the cell

Life can be seen as a constant flow of energy manipulated through chemical reactions to carry out the work necessary to grow, maintain and replicate its forms. The sum of all these chemical reactions is termed metabolism. The laws of thermodynamics govern these energy transformations (see Section 3.1.1). The First Law of Thermodynamics simply states that energy cannot be created or destroyed – it can only undergo conversion from one form to another. The Second Law of Thermodynamics states that disorder in the universe is constantly increasing: energy changes naturally to less organised and useful forms. Thus with every transfer of energy, some energy is lost as heat, a poor quality of energy for biological systems although valuable to warm-blooded species. Thus, although the total amount of energy in a closed system does not change, the amount of useful energy declines as progressively more energy degrades to heat in the reactions. Fortunately, although the universe is a closed system, the biosphere is not and it receives constant supplies of energy from the sun to make up for that lost during life processes. This is where the vital role of photosynthesis to an ecosystem is evident (Box 1.3).

The cell's role is to acquire energy and materials and to use them to maintain itself and to reproduce. Energy stored in chemical bonds can be transferred to new, less energetic bonds, releasing energy in the process or, by adding energy, the potential energy of the new compound can be increased.

The compartmentalisation provided by cell organisation allowed cells to develop areas of enzyme specialisation so that, within each organelle, chemical reactions become organised into metabolic pathways. Some are degradative, breaking down complex chemicals with the release of energy, while others are largely synthetic, requiring the input of energy for the construction of more complex molecules or structures. These two forms of reaction are not exclusive and may interact. Each reaction begins with a substrate acted upon through the mediation of enzymes to produce an end product; many chemical intermediates may be formed during the process.

Enzymes are organic catalysts based on protein molecules that speed up chemical reactions. All enzymes are very specific in their action and no reactions can occur in a cell unless the correct enzyme is present and active. Any process that disrupts enzyme function can have a dramatic, sometimes lethal, effect upon cell activities. Enzymatic reactions may proceed very rapidly, e.g. the breakdown of hydrogen peroxide into water and oxygen may occur 6×10^5 times every second when the enzyme catalase is present. A variety of factors determine how quickly enzymes work, including:

- concentration of the reacting substance (called the substrate)
- temperature
- pH
- enzyme concentration.

Cells have built-in mechanisms to manipulate and control both enzyme concentration and activity.

Many of the chemical reactions that take place in both prokaryotic and eukaryotic cells require an energy input supplied by the molecule ATP (adenosine triphosphate). This is a high-energy compound whose phosphate groups are easily removed in chemical reactions, releasing about 7.3 kcal per mole of ATP each time. ATP supplies energy for three main needs:

- **Chemical work**, supplying the energy needed to synthesise complex molecules needed by the cell
- **Transportation work**, providing energy needed to pump substances across the plasma membrane
- **Mechanical work**, providing energy needed for chromosome movement, cilia and flagellar movement and muscle contraction.

Energy is released by ATP being converted to ADP (adenosine diphosphate) plus phosphate, releasing energy in the process. ATP is reformed using energy obtained from glycolysis by a coupling reaction between the ADP and a phosphate group.

Box 1.3

Photosynthesis – the foundation of the biosphere

About 3 billion years ago, some primitive organisms evolved the ability to capture photons from sunlight and, through a series of molecular activations, synthesised organic molecules. About 1.5 to 2 billion years ago, further evolution resulted in a modified form of photosynthesis which released oxygen as a by-product. This eventually changed the earth's atmosphere from a reducing one to an oxygen-rich one, allowing the evolution of highly efficient aerobic respiration pathways for the extraction of energy from food. Now the two principal biochemical pathways, photosynthesis and respiration, were interlinked because of their by-products. Photosynthesis produces food and oxygen while respiration uses the food to produce usable energy and releases carbon dioxide for further photosynthesis.

The basic formula for photosynthesis is:

$$6CO_2 + 12H_2O + \text{light energy} \rightarrow 6O_2 + C_6H_{12}O_6 + 6H_2O$$

This apparently simple equation hides the fact that it is carried out by a complex series of reactions. A special combination of enzymes together with special elements of cell architecture combine to harvest light energy: the importance of water in these processes is obvious.

The vital substance in photosynthesis is the green pigment, *chlorophyll*, which captures sunlight and converts it into chemical energy: there are several varieties of chlorophyll but the most widespread is chlorophyll a. Not all wavelengths of light can be captured by chlorophyll a, however. The action spectrum (Figure B1.1) reveals that other forms of chlorophyll and some carotenoids assist in broadening the spectrum of photosynthetically active radiation (PAR). This, however, remains dominantly in the red region of the visible spectrum. The pigments are typically contained in a special cell organelle, the *chloroplast*, a membrane-bounded organelle with its own genetic material. This feature has led to the theory that it originated from an ancient, endosymbiotic, photosynthetic bacterium. Inside, it contains a highly organised series of flattened, interconnected compartments known as thylakoids, on which the key biochemical reactions take place. The biochemical reactions are complex and give rise to three main biochemical adaptations to the environmental conditions, referred to as C3, C4 and CAM metabolisms (see Box 4.5).

An important feature of photosynthesis is that for every 100 units of light energy falling on a plant, only between 2 and 5 units are converted into plant material. Thus plants are not very efficient but because their energy resource, sunlight, is inexhaustible this of little significance under most conditions. Figure B1.2 summarises the key features of photosynthetic production.

Box 1.3 continued

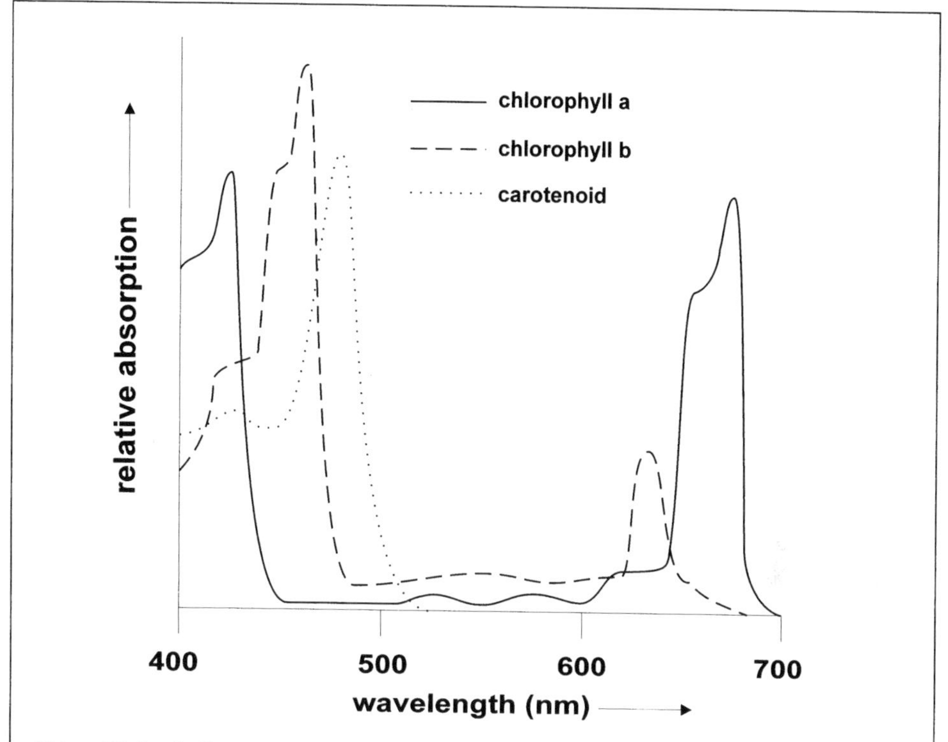

Figure B1.1 ***Action spectrum for chlorophylls a and b and a carotenoid***

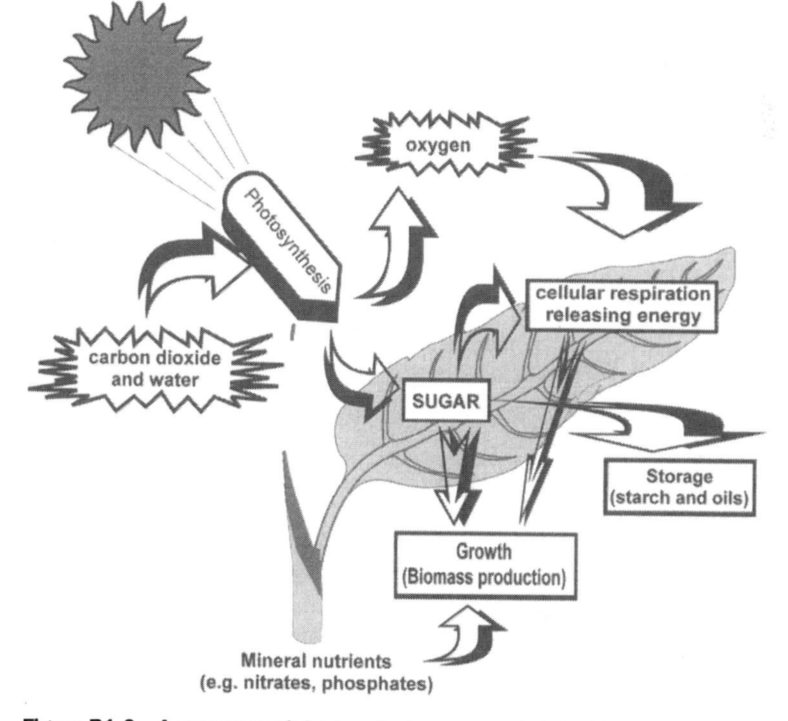

Figure B1.2 ***A summary of the key features associated with photosynthetic production***

1.1.3 The materials of life

Only eleven of the ninety-two elements found naturally in the earth's crust are common in living organisms (Table 1.3) but many others are important despite occurring only in trace quantities. More than 97 per cent of atoms in the human body are nitrogen, oxygen, carbon or hydrogen. About 90 per cent of the atoms common in organisms are hydrogen and oxygen, reflecting the significance of water (H_2O) in living systems. The single most important property of water is its ability to form weak chemical associations with only 5–10 per cent of the strength of normal covalent chemical bonds. This results in water possessing special properties that are vital to life (Box 1.4).

Table 1.3 ***The main life elements and their relative abundances in the Earth's crust and the human body***

Element	*Chemical symbol*	*% by weight in Earth's crust*	*% by weight of body*	*Main role/s*
Oxygen	O	46.6	65.0	cellular respiration/component of water
Calcium	Ca	3.6	1.5	skeletal agent/cell adhesion/muscle contraction
Sodium	Na	2.8	0.2	main positive ion bathing cells/nerve function
Potassium	K	2.6	0.4	main positive ion in cells/nerve function
Magnesium	Mg	2.1	0.1	component of many enzymes
Hydrogen	H	0.14	9.5	electron carrier/component of water and most organic molecules
Phosphorus	P	0.07	1.0	nucleic acids/important in energy transfer (ATP system)
Carbon	C	0.03	18.5	basis of all organic molecules
Sulphur	S	0.03	0.3	component of most proteins
Chlorine	Cl	0.01	0.2	main negative ion bathing cells
Nitrogen	N	<0.01	3.3	component of all proteins and nucleic acids

The other important molecules in living material are gigantic compared to the water molecule and are called macromolecules. There are four general categories of macromolecules: carbohydrates, lipids, proteins and nucleic acids. These are the basic building blocks used to construct cells, the basic operational units from which the great diversity of life has evolved. These molecules contain *functional groups*, i.e. subgroups of atoms with specific chemical properties that largely determine the types of reactions that the molecule is capable of participating in (Table 1.4). Most such reactions involve the transfer of functional groups between molecules or the breaking of carbon–carbon bonds. Macromolecules are frequently polymers, i.e. large molecules built up as a long chain of similar subunits. The most familiar are the nucleic acids, long-chain polymers made up from subunits called nucleotides (see Section 1.1.3.4), whose sequences carry the genetic information essential for reproduction.

Box 1.4

Water – a unique substance and the elixir of life

This seemingly simple molecule has many surprising properties that are important to life, from cellular chemical level to environmental level. Water covers three-quarters of the Earth's surface and comprises about two-thirds of most organisms. It is remarkable because it can exist in all three physical forms at the normal environmental temperatures found on Earth – solid (ice), liquid (water) and gas (water vapour). This is a consequence of water being a polar molecule, i.e. it has a minute electrical charge, with the resultant ability to form hydrogen bonds. They are very weak and transient (each lasting only 10^{-11} seconds) inter-molecular bonds whose cumulative effects have a major influence on the properties of water. Without these bonds, water would boil at –80°C and freeze at –100°C, making life as we know it impossible. The main biologically important properties are:

- Water is a universal solvent and facilitates chemical reactions between polar molecules and ions (hydrophilic molecules) which will dissolve in the water thus being brought together. Non-polar and non-ionised molecules are hydrophobic ('water-hating'; repelled by water molecules) and are not readily soluble.
- Water molecules are cohesive, clinging together because of hydrogen bonding. When in contact with other polar surfaces, this attraction is called adhesion. This results in surface tension effects and capillary action, both of which are important environmental and biological phenomena, e.g. allowing development of biological transport systems.
- The temperature changes of liquid water in the environment and in the body are 'damped', i.e. reduced in magnitude. This is because water has a high specific heat, a measure of the amount of heat required to raise the temperature of a unit mass, because of the hydrogen bonding. This means that water heats up more slowly than almost any other compound and will retain that heat longer. Thus water protects organisms from rapid temperature changes, both internally and externally, e.g. the heat generated by metabolism would 'cook' cells if it were not for the water absorbing most of this heat.
- Water resists changes of state because it has a high heat of vaporisation – i.e. it requires the input of relatively large amounts of energy to convert liquid water to vapour. Thus the evaporation of water from surfaces produces significant cooling (evaporative cooling) and many higher organisms dispose of excess heat by this process (e.g. sweating).
- Water is most dense at 4°C, well above its freezing point of 0°C: this is due to the hydrogen bonds forming a relatively open molecular lattice as water cools below 4°C. Therefore, ice is lighter than liquid water and bodies of water freeze from the top downwards. The ice layer on the surface then acts as an insulator to prevent further freezing of the deeper layers. This property has been vital for the survival of aquatic faunas during the winter in high-latitude freshwater systems.
- Water tends to ionise to form hydrogen ions (H^+) and hydroxyl ions (OH^-). The concentration of hydrogen ions is reflected in the pH scale used to describe the acidity (excess H^+ relative to pure water) or alkalinity (lowered H^+ relative to pure water) of a liquid. Pure water has a pH of 7, the molar concentration of hydrogen ions being 10^{-7} moles per litre and pH being $-\log [H^+]$. Each unit change in pH represents a tenfold change in the hydrogen ion concentration.

Table 1.4 ***The main functional groups in chemicals***

Functional group	*Formula*	*Compounds*
Hydroxyl Group	-- OH	Simple and complex alcohols
Carbonyl Group	-- C -- (=O)	Aldehydes and ketones
Carboxyl Group	–C(=O)OH	Carboxylic acids
Amino Group	–N(H)H	Amines
Sulphydral	-- S --H	Thiols
Phosphate	-- O -- P(OH)(=O) -- OH	Organic phosphates

1.1.3.1 Carbohydrates

Carbohydrates vary in size from simple, small molecules to long polymers and function both as structural elements and as energy storage materials. Sugars (general formula $(CH_2O)_n$) are important energy-storage chemicals, the energy being released when the abundant C–H bonds are broken by enzymatic action. The most important quick-release energy-storage form is the monosaccharide glucose, having six carbon atoms and seven C–H bonds and the formula $C_6H_{12}O_6$ (Figure 1.2). This diagram shows that other sugars can occur with the same chemical formula. These are **isomers**, chemical compounds with the same molecular formulae but different molecular structures or different spatial arrangements of their atoms. Thus, fructose has the double bonded oxygen attached to an internal carbon atom rather than a terminal one (a structural isomer) while galactose is a mirror image of glucose (a stereoisomer). These apparently small differences are of considerable significance for living organisms since the various isomers may have significantly different biochemical behaviours.

Many organisms use various disaccharides, formed by linking two monosaccharide molecules, as reservoirs of energy. They cannot be metabolised to release the stored energy until they are broken down into glucose units by the appropriate enzyme. Storage occurs by further converting the disaccharides into insoluble, long polymers called polysaccharides: those formed from glucose are starches and are important plant products, e.g. potato starch. Animals construct the insoluble polysaccharide glycogen as their principal form of glucose store. The polysaccharide cellulose is the main structural constituent of plant cell walls and is very difficult to break down. By adding a nitrogen group to the glucose units of cellulose, organisms such as insects and many fungi produce chitin, another highly resistant structural material (Box 1.5).

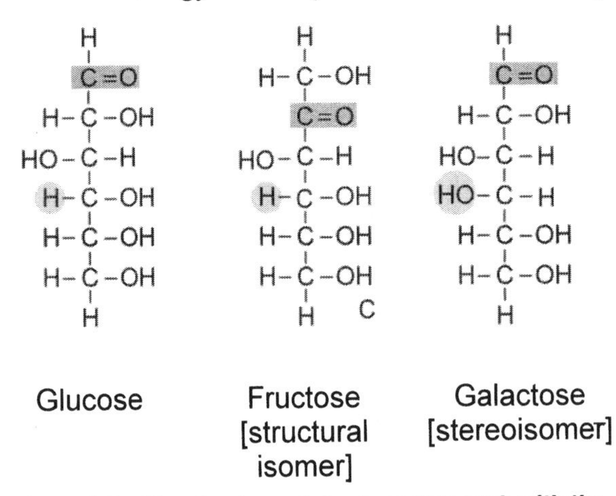

Figure 1.2 ***The structure of simple sugars each with the chemical formula $C_6H_{12}O_6$ and showing the two isomeric forms of glucose***

Box 1.5

Cellulose and chitin – nature's tough materials

Cellulose is a polymer built from glucose (sugar) units joined differently from those in the easily broken down storage polymers also built from glucose units, namely starch and glycogen. The orientations of the chemical bonds in cellulose cause the polymer to be straight and fibrous, an ideal structural material. The long unbranched polymers of cellulose are held together in rigid cross-linked groups by hydrogen bonding to form microfibrils, several of which together form a fibril that has great tensile strength. Groups of fibrils in variously orientated layers develop in plant cell walls which remain fully permeable to water and solutes. Cellulose makes up between 20 and 40 per cent of plant cell walls and about 50 per cent of the carbon found in plants is cellulose. Although chemically similar to starch (Figure B1.3), the starch-degrading enzymes that occur in most organisms cannot digest cellulose.

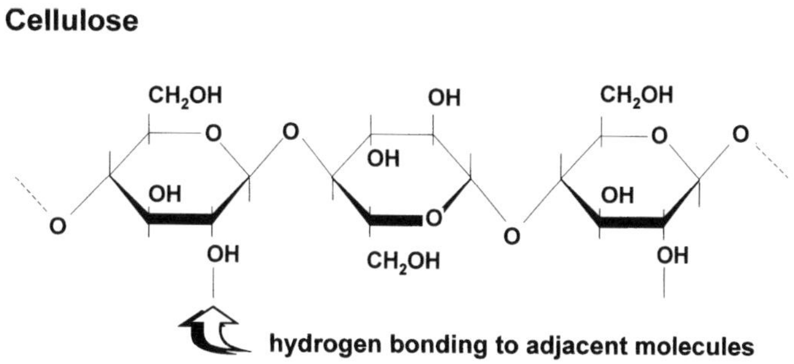

Figure B1.3 ***The basic structure of the subunits of cellulose***

The main structural material in arthropods and some fungi is chitin, a modified form of cellulose in which the hydroxyl group at carbon atom 2 is replaced by $-NH.CO.CH_3$ (Figure B1.4). This renders it a tough and resistant surface material widely used in exoskeletons. Like cellulose, it cannot be digested by most organisms.

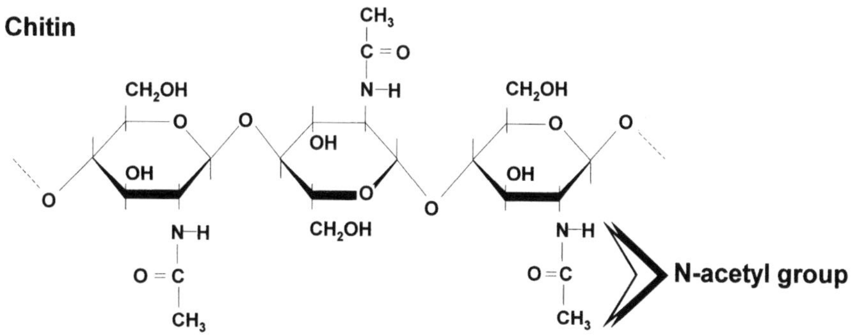

Figure B1.4 ***The basic structure of the subunits of chitin***

Cellulose is the most abundant organic molecule on Earth and humans use it in many ways. Wood contains a high percentage of cellulose and is widely used for building and making paper; cotton fibre is almost pure cellulose. The possible uses for chitin are currently being investigated but it is already used as a thread for surgical sutures.

Box 1.5 continued

Few organisms are able to digest cellulose or chitin. Those that can possess enzymes termed cellulases and include some protozoa and fungi, a few insects, earthworms, wood-boring molluscs and bacteria. Some plant-eating organisms such as ruminant mammals and termites digest cellulose by utilising colonies of cellulose-digesting protozoa and bacteria cultured within their digestive systems (see Box 4.8). Humans do not possess such colonies and cannot digest cellulose but it remains an important part of our diet by forming the roughage essential to normal functioning of the digestive system.

The abundance of cellulose and its relatively slow rate of degradation have important ecological implications since much carbon becomes 'locked up' for long periods as cellulose. The significance for the carbon cycle (Section 3.3.2) is clear.

1.1.3.2 Lipids

There are many different kinds of lipids, all of which are non-polar and are, therefore, insoluble in water. The main biologically active ones include fats, oils, waxes, phospholipids, steroids, prostaglandins and terpenes. Each fat molecule comprises two kinds of subunits: glycerol, a three-carbon alcohol, and fatty acids, long hydrocarbon chains ending in a carboxyl (–COOH) group. Three fatty acids attach to each glycerol molecule (Figure 1.3) and the resulting molecule is a triglyceride; the component fatty acids may differ markedly from each other, leading to different properties. When all of the internal carbon atoms have hydrogen side groups, the fatty acids are termed 'saturated'. If, however, double bonds are present between one or more pairs of successive carbon atoms, they are unsaturated. If more than one double bond is present, they are termed 'polyunsaturated' and have low melting points; such liquid fats are called oils. Many plant fatty acids, e.g. linolenic acid (linseed oil), are unsaturated whilst animal fats are usually saturated. Fats are very efficient energy-storage molecules because of their abundance of C–H bonds. Most fats contain more than forty carbon atoms and saturated fats are more effective stores of energy than unsaturated ones. Typically, fats store about twice the energy of the equivalent amount of carbohydrate.

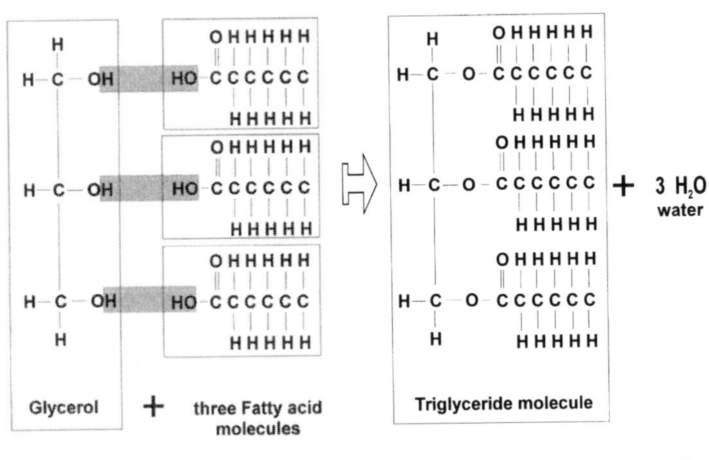

Figure 1.3 ***The construction of a fat molecule by a dehydration reaction***

Phospholipid molecules form the vital cell membranes and have a polar group at one end and a long, non-polar tail. In water, this results in the formation of a lipid bilayer (Figure 1.4) providing the basic framework of all biological membranes and in

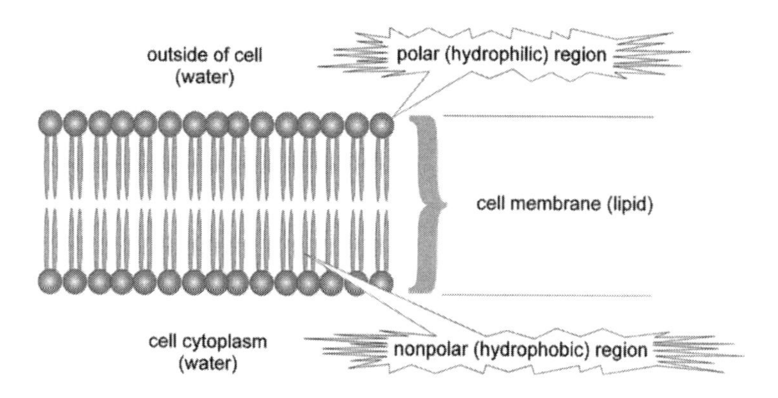

Figure 1.4 ***The lipid bi-layer construction of the cell's plasma membrane***

particular the plasma membrane. Such membranes are vital in the control of the relationship between the cell and its environment.

Terpenes are long-chain lipids which form many of the biologically important pigments such as the photosynthetic pigment chlorophyll. Other important lipids include steroids such as cholesterol and the sex hormones, and prostaglandins, modified fatty acids found in vertebrates where they act as local chemical messengers.

1.1.3.3 Proteins

Despite the diverse functions of proteins (Table 1.5), all have the same basic structure of a long polymer chain (a polypeptide) of amino acid subunits joined end to end and linked by peptide bonds. An amino acid contains an amino group (–NH_2), a carboxyl group (–COOH) and a hydrogen atom, all bonded to a central carbon atom (Figure 1.5). Although many different amino acids occur in nature, there are only twenty in organisms. These fall into five chemical classes, each having distinctive chemical properties due to the diversity of the side groups. However, proteins can be formed from any sequence of amino acids and thus a protein composed of 100 amino acids can have 20^{100} different sequences. This permits great diversity in the kinds of proteins constructed within cells. The chemistry of proteins is complex and depends upon molecular shape as well as the chemical content. Because of hydrogen-bonding between adjacent amino acids, polypeptide chains tend to fold spontaneously into sheets or wrap into coils: interactions with water can result in further folding into globular shapes. The results of all these interactions are subtle or profound differences between proteins even when there is only a single amino-acid difference. It is because of this diversity of specific shapes that proteins are such effective biological catalysts.

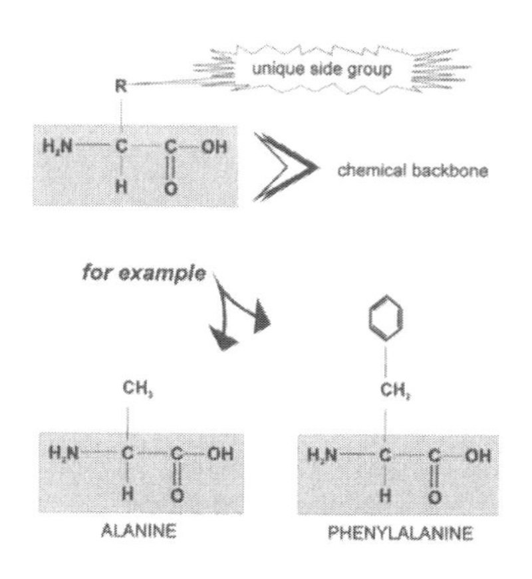

Figure 1.5 ***The structural plan of amino acids***

1.1.3.4 Nucleic acids

Nucleic acids are very large molecules that act as the blueprints and information storage devices for cells, in the same way that disks or tapes store programs and

Table 1.5 ***Some of the many functions of proteins***

Function	*Type of protein*	*Examples*	*Use*
Structural	Fibres	Collagen	cartilage
		Keratin	hair and nails
		Fibrin	blood clotting
Metabolism	Enzymes	Lysosomes	cleave polysaccharides
		Proteases	break down proteins
		Polymerases	create nucleic acids
		Kinases	phosphorylate sugars/proteins
Membrane transport	Channels	Na^+/ K^+ pump	excitable membranes
Cell recognition	Cell surface antigens	ABO blood groups	identifies red blood cells
Regulation of gene action	Repressors	Lac repressor	regulates transcription
Regulation of body functions	Hormones	Insulin	controls blood glucose levels
		Oxytocin	regulates milk production
Transport through body	Globins	Haemoglobin	carries O_2 and CO_2 in blood
		Cytochromes	electron transport
Storage	Ion-binding	Ferritin	stores iron, especially in spleen
		Casein	stores ions in milk
Contraction	Muscle	Actin	contraction of muscle fibres
		Myosin	contraction of muscle fibres
Defence	Immunoglobulins	Antibodies	mark foreign proteins for elimination
	Toxins	Snake venom	blocks nerve function

Source: Adapted from Raven and Johnson 1992.

information for computers. They come in two varieties: deoxyribonucleic acid (DNA) and ribonucleic acid (RNA). Unique among macromolecules, DNA is able to produce exact copies of itself, enabling the passing of information specifying the amino-acid sequence of its proteins to descendants and forming the basis of hereditary material. The information is stored as a code represented by the sequence of four subunits called nucleotides. Cells use RNA to read each DNA-coded information unit (gene) and then use this information to produce the appropriate protein.

Most organisms have their DNA as a double-helix structure composed of sequences of four nucleotides. Each nucleotide is a molecule composed of three smaller molecules, a five-carbon sugar (deoxyribose in DNA), a phosphate group (PO_4) and an organic nitrogen-containing base (Figure 1.6). The sugar and phosphate groups form a backbone for the support of the bases that occur in four distinct types that form the 'letters' of the genetic code. Two of these are the larger purines, adenine (A) and guaninė (G) whilst the other two are the smaller pyrimidenes, thymine (T) and cytosine (C). To fit into the double-helix chain, a large nucleotide must always be opposite a smaller one to which it becomes attached by a hydrogen bond. Because of this and other chemical parameters, A and T can only pair with each other, as can G and C. This results in one strand of DNA fully specifying the other strand and the strands are complementary (Figure 1.6). This is the basis for the precise replication of nucleic acid sequences (genes) during reproduction, the process which underlies heredity.

RNA is similar to DNA but with two significant chemical differences:

1 the sugar is ribose rather than deoxyribose and

2 the nucleotide thymine (T) is replaced in the code by uracil (U).

This allows the cell to distinguish between the RNA copy and the original, aided by the fact that DNA always forms a double helix while RNA is usually single-stranded.

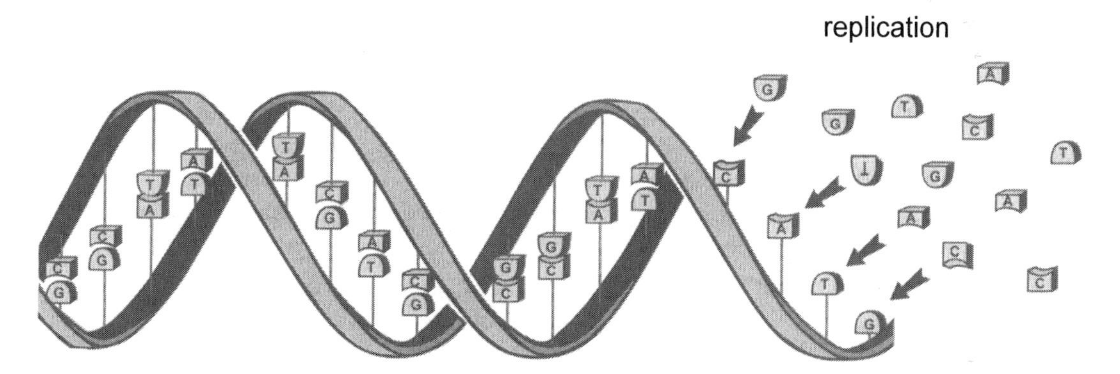

Figure 1.6 ***The helical structure of DNA showing the principle of base pair replication***

1.2 The diversity of life

The diversity of living forms is remarkable. It has been estimated that there are at least 10 million kinds of organisms on Earth although biologists do not know with any degree of certainty how many actually exist: some estimates are as high as 100 million. Only about 1.5 million have been named so far. Such biodiversity consists of various levels of organisation and this section introduces the key components used in the description of such diversity.

1.2.1 Levels of organisation

Living organisms can be considered at various levels of organisation, dealt with in detail later in this book. Despite the unifying features of cells and subcellular structures/ mechanisms discussed in Section 1.1.2, the biosphere contains an amazing variety of shapes, sizes and life histories.

- The basic unit is the *individual*, which can be a single-cell protistan or a large, complex multicellular organism such as a blue whale, the Earth's largest-ever animal, or the coastal redwood, the largest plant. The concept of the individual is valid for most types of organism but note that organisms such as corals form colonies in which functionally discrete individuals often cannot be distinguished (Figure 1.7). Discrete organisms are termed *unitary* organisms and most adult members of the population are of similar size and shape: thus all humans have only two legs while all insects have six legs. However, the number of leaves on a tree or the number of polyps in a coral is not constant. They are *modular* organisms made up of many repeatable units (modules), each of which is the equivalent of a cloned individual. Each module has its own cycle of birth and death, largely independent of the whole structure, which is referred to as a colony. Each module can be studied as if it were an individual and studies are often concerned primarily with the number, size and shape of modules rather than with the number of genetically distinct individuals, known as **genets**. Most plants are modular and there are many important groups of modular protistans, fungi and animals such as sponges and corals.

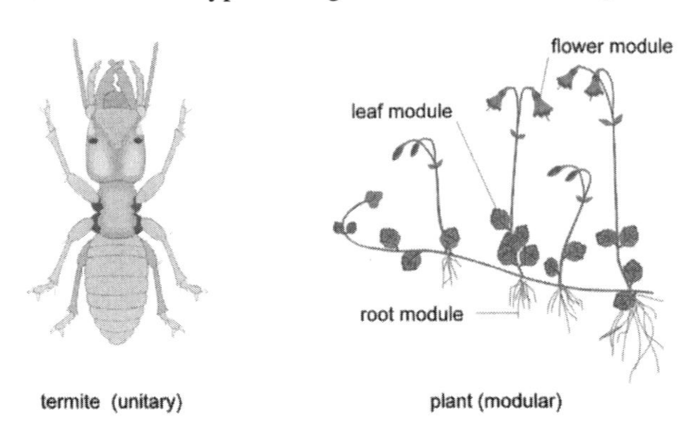

Figure 1.7 ***Examples of a unitary and a modular organism***

- A group of individuals of the same species living within a defined area is a *population*, the major focus of interest for many ecologists and evolutionary biologists. Population biologists study parameters such as population size or abundance, how size changes with time, the factors determining population size, and how population characteristics vary over time. The study of the biology of single species is **autecology**. Understanding the factors which influence these parameters is a key area of ecological study and has major significance for applied topics such as natural resource management, conservation and biological control of pest populations. Populations rarely exist in isolation, normally coexisting with populations of other species to form ecological communities.
- The concept of a *community* is rather abstract, having originated from studies of plant aggregations. It was extended later to collections of any species that occur together in a common environment or habitat and which are biologically integrated or interact as a group. Most communities comprise a diverse mixture of groups from any or all of the kingdoms of organisms and occur repeatedly in similar habitats. They often take their name from the physical features of the habitat such as lake communities and sand dune communities. Others are named after the dominant life forms in the association, such as deciduous oak wood, coral reef and sphagnum bog communities. Community boundaries are often poorly defined and it is important to remember that the biotic environment is highly variable both

spatially and temporally. The organisation and functioning of communities is another important area of environmental study.

- The term *ecosystem*, although widely used, is similarly an abstract concept since it refers to the complex of interactions between collections of communities and their physical environment, without there being distinct boundaries or defined scales. Thus, reference can be made either to the marine ecosystem or to the rock pool ecosystem, depending upon the level at which the system is considered. The important concept is that, within an ecosystem, the essential material components on which the constituent populations and communities depend are recycled while energy is regulated and transferred. Evolution over millions of years has led to the creation of diverse terrestrial, freshwater and marine ecosystems.

To understand the nature and extent of diversity, the concept of the species, the basic taxonomic unit, needs to be introduced.

1.2.2 The species concept

Taxonomists, whose role is to describe and classify living organisms, divide living forms into discrete units called **species**. The study of natural populations reveals that they tend to occur in units which are distinctly different in form (**phenotype**), ecology, genetic composition (**genotype**) and evolutionary history, but within which individuals resemble each other. However, the definition of a species remains a matter of some debate.

Linnaeus separated species on the basis of morphology (form and structure) and Darwin first recognised that species are related by common ancestry. Mayr's (1970) definition of a biological species – 'Species are groups of actually or potentially interbreeding natural populations which are reproductively isolated from other such groups' – emphasises the ability to reproduce successfully, separate species being prevented from merging with each other by a variety of isolating mechanisms which form barriers to gene flow. Isolating mechanisms comprise any structural, functional or behavioural characteristics that prevent successful reproduction and include:

- **Habitat isolation**: occupying different habitats within the same geographical range (adults are unlikely to meet)
- **Temporal (seasonal) isolation**: reproducing and/or maturing at different times of year
- **Behavioural isolation**: having different courtship patterns involving visual, auditory, tactile and olfactory stimuli
- **Mechanical isolation**: development of incompatible reproductive structures
- **Gamete isolation**: making male and female gametes (sex cells) incompatible even if gamete transfer takes place
- **Hybrid failure**: hybrid inviability, where embryos fail to develop to maturity; hybrid sterility, where hybrids fail to produce functioning sex cells.

The major problem with the *Biological Species Concept* is that it is not applicable to species that reproduce asexually or that self-fertilise (see Section 5.5) and it cannot be applied to fossil forms. Thus all extinct organisms and the large number of organisms which are not obligate cross-fertilising sexual forms are excluded from this definition. However, these organisms can be divided pragmatically into species on the basis of the same pattern of phenotypic (observable features) similarity and discontinuity between species used for living sexual species.

Two alternative species definitions have been proposed, the *Evolutionary Species Concept* and the *Recognition Species Concept*. The Evolutionary Species Concept considers a species to comprise all those individuals that share a common evolutionary history. It is useful because it is similar to the practice used by most taxonomists and palaeontologists but it has its own problems. These include judging what comprises a common evolutionary history and describing how species are maintained or evolved.

The Recognition Species Concept developed during the 1980s considers a species to comprise that inclusive population of individual biparental organisms which share a common fertilisation system. However, this suffers from the same basic problems as the Biological Species Concept, despite placing its emphasis onto the mechanisms keeping a species together rather than those separating them.

Generally, therefore, species are separated on the basis of similarity of phenotype and genotype within the species and differences between the species. Most species are defined on the basis of morphological differences but modern taxonomy has begun to use molecular genetic techniques to investigate more detailed genetic relationships. These have revealed the existence of *sibling species*, groups of morphologically identical individuals that are actually two different species, distinguishable behaviourally, genetically and/or chromosomally and with little hybridisation between them. In practice, despite its limitations, the Biological Species definition is most widely used by those working on sexually reproducing species.

1.2.3 Classification and the binomial system

To describe the amazing diversity of life on Earth, science has evolved a formal method for the naming of species. The binomial system derives from the work of Linnaeus; each species name starts with the generic name followed by the specific epithet. Both are latinised and written in italics but only the genus is capitalised. Scientific names are the same throughout the world and provide a standard way of referring to organisms while common names often vary with the part of the world in which they are used. Thus the sperm whale (also known as the cachelot) is properly known as *Physeter catedon*: this shows that it belongs to the genus (plural genera) *Physeter* and the species name is *catedon*. Specific epithets can only be used in association with a generic name and the same specific epithets may, therefore, be used in different genera. The specific epithet is often chosen to be descriptive of the organism, of its geographical locality, or is named after its discoverer. The scientific name is often followed by a person's name or abbreviation of a name, identifying the scientist who first described that species. The naming of species is subject to very precise rules and is subject to international regulatory committees. So far, about one and a half million species have been named but it is generally considered that at least 10 million species exist.

Table 1.6 ***Hierarchy of taxa: examples of a bacterium, a plant and an animal***

Common name	*Pseudomonas*	*English oak*	*Honey bee*
Kingdom	Monera	Plantae	Animalia
Phylum/Division	Gracilicutes	Anthophyta	Arthropoda
Class	Scotobacteria	Dicotyledonae	Insecta
Order	Pseudomonadales	Fagales	Hymenoptera
Family	Pseudomonadaceae	Fagaceae	Apidae
Genus	*Pseudomonas*	*Quercus*	*Apis*
Species	*P. aeruginosa*	*Q. robur*	*A. mellifera*

Having given each species a name, taxonomists (the scientists who study and classify organisms) then group organisms into a hierarchical system composed of discrete taxonomic ranks or levels, known as taxa (Table 1.6). There are two approaches to this classification. The *phenetic* (numerical taxonomy) approach is based solely on classifying by morphological similarity while the *phylogenetic* (or **cladistic**) approach is based upon classification by evolutionary relationships (Figure 1.8). Both systems have their advocates and both influence modern taxonomic opinion since information about the order of evolutionary divergence and the magnitude of the changes that have occurred are important.

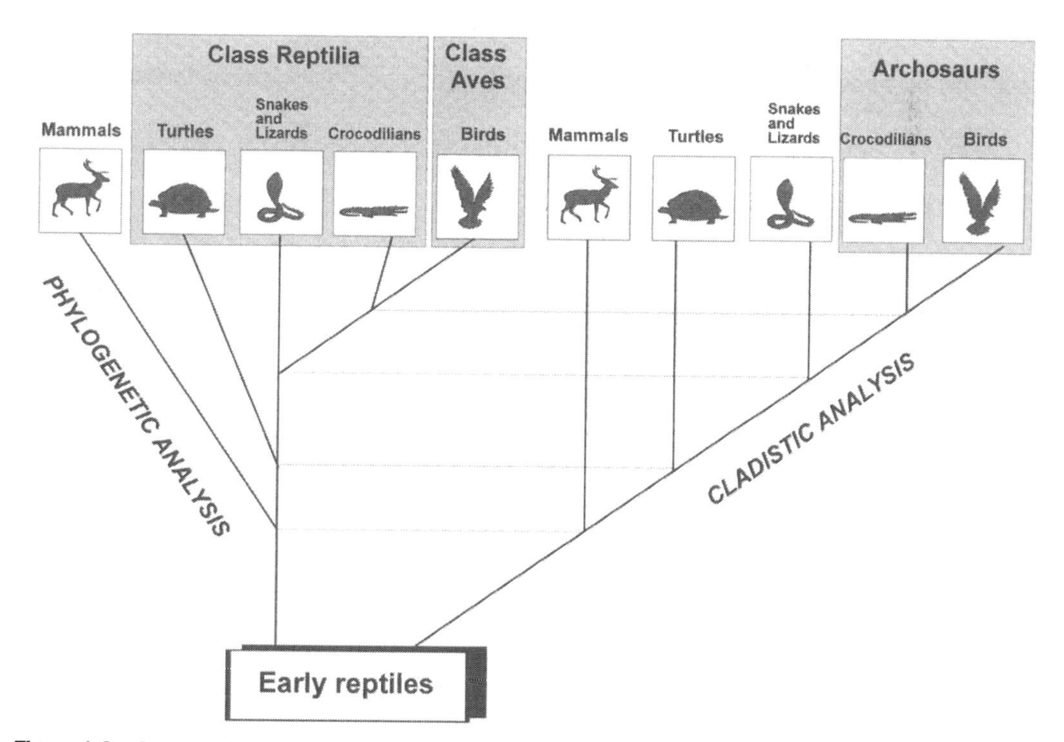

Figure 1.8 ***A comparison of the cladistic and a phylogenetic classification of some vertebrate groups. Note in particular the relative position of the birds (after Raven and Johnson 1992)***

The classification of organisms into kingdoms is somewhat arbitrary, most scientists recognising five until recently when the bacteria were separated into two distinct kingdoms (Figure 1.9).

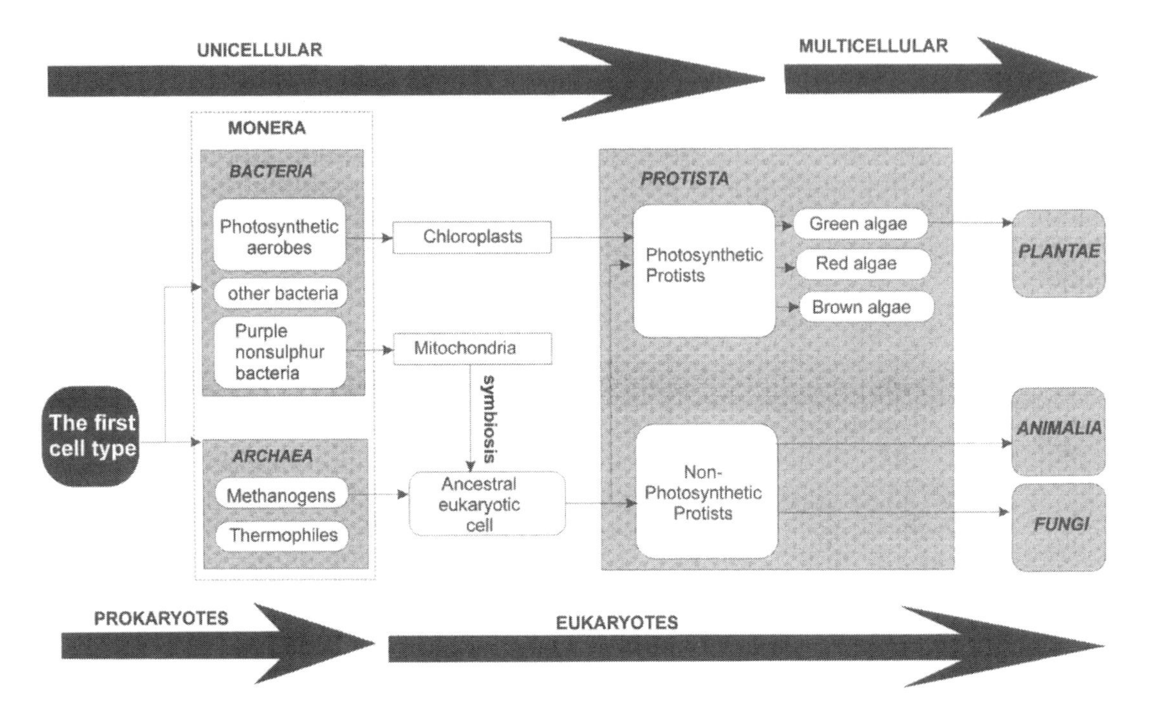

Figure 1.9 ***The six kingdoms and their evolutionary relationships (modified from Raven and Johnson 1992)***

1.2.4 The six kingdoms

Leaving aside the question of the viruses (Box 1.6), sometimes placed in a separate kingdom (Virales), six kingdoms of living organisms are recognised. The most significant distinction between organisms is that separating the prokaryotes and the eukaryotes. The prokaryotes existed for at least two and a half billion years before the appearance of eukaryotes and were until recently all placed in a kingdom called the Monera. They all lack membrane-bound organelles or microtubules and their DNA is not associated with chromosomes nor is it located in a nucleus. However, the Monera was divided recently into two separate kingdoms, the Archaea (formerly Archaebacteria) and the Bacteria (formerly Eubacteria).

The *Archaea* (Table 1.7) possess a unique metabolism and genetic machinery. Many are chemosynthetic, i.e. they are able to utilise chemical energy from sulphur compounds as an energy source rather than sunlight; they also have unusual cell walls and genetic machinery.

Bacteria (Table 1.8) are typically photosynthetic or heterotrophic and lack the unique features of the Archaea.

Box 1.6

Viruses and their relatives: borderline cases

The viruses present a particularly difficult classification problem since they are strictly non-living forms and cannot function outside a host cell. However, they are infectious agents capable of replication by using the machinery of host cells to manufacture new viral materials. They have been found in almost every kind of organism and are usually highly host-specific. The smallest are only about 17 nanometres in diameter, the largest about 1,000 nanometres in their maximum dimension, most requiring electron microscopes for their study. They are composed mainly of a nucleic acid core surrounded by a protein coat (capsid) which is itself often surrounded by a lipid-rich protein envelope. They are much smaller than bacteria. Many plant viruses such as tobacco mosaic virus (TMV) have a core composed of RNA while others contain DNA. Viruses frequently disrupt the normal functioning of the host cell and are responsible for causing disease. They are important disease-causing agents in plants which can be the cause of significant economic damage in horticulture and agronomy. Animal infections can be equally damaging for agriculture and include diseases such as cowpox and the rabbit myxomatosis virus. In humans, viruses cause diverse diseases such as measles, influenza, polio, rabies and AIDS.

Bacteriophages are a form of virus which exclusively infect bacteria and are characterised by having a tail with 'legs' (tail fibres), used to attach to the bacterial cell. They are more complex than viruses infecting plant and animal cells.

Viroids are the smallest known pathogens, forming a third type of infectious agent. They resemble viruses but are not strictly viruses, being only small, circular molecules of RNA lacking a capsid and any proteins. So far, they appear to be confined to plants, causing about a dozen known plant diseases.

Prions are infectious agents composed only of protein. They cause important diseases such as scrapie in sheep, bovine spongiform encephalopathy (BSE) in cattle and Creutzfeldt-Jacob disease (CJD) in humans. Although both viroids and prions are very simple, both are infective and, like viruses, reproduce within host cells.

Eukaryotes are distinctly different from prokaryotes, as described earlier (Section 1.1.2), and exhibit true sexual reproduction and integrated multicellularity, facilitating the diversity of form and function found in today's world. Despite this, they are much less diverse metabolically than the Bacteria.

The diverse array of single-celled eukaryote phyla are placed in the kingdom *Protista* (Table 1.9) which today includes organisms which were once considered to be either algae (photosynthetic protists) or animals (heterotrophic protists = protozoa). Although these terms are still used, this is no longer considered to reflect the true relationships between protist groups. They are now recognised as being so diverse that taxonomists are constantly proposing new classifications of the component groups.

Table 1.7 ***The main taxa of the Archaea***

Group	*Key features*
Methanogens	manufacture methane, are anaerobic, found in swamps and marshes
Thermoacidophiles	favour hot and acidic habitats such as hot springs
Halophiles	Archaea which require saltwater conditions

Table 1.8 ***The main taxa of the Bacteria***

Group	*Examples*	*Key features*
Actinomycetes	*Streptomyces, Actinomyces*	Gram-positive bacteria often forming branching filaments; produce many antibiotics such as streptomycin; common soil bacteria
Chemoautotrophs	sulphur bacteria, *Nitrobacter, Nitrosomonas*	Chemosynthetic group usually utilising hydrogen sulphide, ammonia or methane; a key group in the nitrogen cycle
Cyanobacteria	*Oscillatoria, Spirulina*	Photosynthetic bacteria common in aquatic habitats; often responsible for toxic blooms in polluted water bodies; includes nitrogen-fixing forms
Enterobacteria	*Escherischia coli, Salmonella, Vibria cholerae*	Gram-negative, rod-shaped bacteria, usually aerobic heterotrophs; many are important disease-causing agents
Gliding and budding bacteria	Myxobacteria, *Chrondromyces*	Gram-negative bacteria which show gliding mobility; some form upright multicellular fruiting structures which develop spores
Pseudomonads	*Pseudomonas*	Gram-negative heterotrophic rods bearing polar flagella, common in soil; includes many important plant pathogens
Rickettsias/Chlamidias	*Rickettsia, Chlamydia*	Gram-negative intracellular parasites with often complex life cycles; include many lethal human diseases
Spirochaetes	*Treponema*	Long, coiled cells with flagella at both ends, common in water

The main features of eukaryotic evolution which today account for the diversity of eukaryote form and function include the development of organelles and multi-cellularity, the evolution of sexual reproduction and the variations in life cycles. True multicellularity, where the activities of cells in contact with each other are co-ordinated, is found only in eukaryotes. Multicellularity evolved independently many times in protists, allowing them to deal with their environments in new ways. They did this by the differentiation of distinct and specialised types of cells, tissues and organs, thus facilitating the separation of functions within the body. This has allowed the development of a complexity impossible for unicellular forms. The development of multicellularity eventually resulted in the development of the remaining three kingdoms, the Plantae, the Fungi and the Animalia. These are sometimes referred to as higher organisms.

Table 1.9 ***The main taxa of the Protista***

Group	*Phyla*	*Key features*
Sarcodina	Rhizopoda	move by pseudopodia; heterotrophic
	Foraminifera	move by protoplasmic streaming; have calcareous shells; heterotrophic
Algae	Chlorophyta	green, photosynthetic, often multicellular, contains chlorophyll a and chlorophyll b
	Rhodophyta	red, photosynthetic, often multicellular; chlorophyll a and a red pigment present
	Phaeophyta	brown, photosynthetic, usually multicellular; chlorophyll a and chlorophyll c present
Diatoms	Bacillariophyta	unicellular, with external shell (test) of silica; photosynthetic
Flagellates	Dinoflagellata	unicells with two locomotory flagellae; photosynthetic
	Zoomastigina	unicells with locomotory flagellae; heterotrophic
	Euglenophyta	unicells with locomotory flagellae; some heterotrophic, some photosynthetic
Ciliates	Ciliophora	unicellular with locomotory cilia and fixed cell shape; heterotrophic, e.g. *Paramecium*
Sporozoans	Sporozoa	unicellular nonmotile parasites; spore-forming, e.g. *Plasmodium*
Moulds (including cellular slime moulds)	Acrasiomycota	heterotrophic with limited mobility; carbohydrate cell walls; colonial aggregations of individual cells related to amoebae
Plasmodial slime moulds	Myxomycota	heterotrophic slime moulds showing streaming as a multinucleate mass of cytoplasm
Water moulds	Oomycota	terrestrial and freshwater rusts and mildews; heterotrophs

The *Plantae* (Table 1.10) evolved from photosynthetic protists and are characterised by photosynthetic nutrition, cell walls made from cellulose and other polysaccharides, lack of mobility and a characteristic life cycle involving an alternation of generations (see later).

The *Fungi* (Table 1.11) are mainly multicellular although the kingdom also contains the important yeasts which are unicellular. Fungi are characteristically filamentous in form, non-photosynthetic (heterotrophic), and environmentally important because of their roles in decomposition and recycling processes.

The *Animalia* (Table 1.12) are all multicellular heterotrophs lacking cell walls and having a nervous system which may be complex. They are frequently highly mobile.

Table 1.10 ***The main taxa of the Plantae***

Phylum/division	*Examples*	*Key features*
Bryophyta	mosses such as *Polytrichum* and *Sphagnum*; liverworts and hornworts (about 16,500 species)	no vascular tissues, no true roots or leaves; live in moist habitats
Coniferophyta	conifers such as redwoods, pines and firs (about 55 species)	flowerless gymnosperms with non-motile sperm; wind pollination and dispersion usual; needle-like or scale-like leaves, most being evergreen; often dominant in cooler climates
Cycadophyta	cycads, sago palms (about 100 species)	very slow growing, palm-like gymnosperms; flagellate sperm reach egg by pollen tube
Gnetophyta	shrub teas and vines (about 70 species)	gymnosperms with non-motile sperm
Ginkgophyta	*Ginkgo* is the only genus	deciduous gymnosperms with motile sperm
Lycophyta	lycopods (about 1000 species)	seedless but diploid vascular plants which look like mosses and occur in similar moist habitats
Pterophyta	ferns and tree-ferns	seedless vascular plants with diploid, dominant sporophyte generation and small but free-living haploid gametophyte generation
Anthophyta	flowering plants including oak, corn, maize and herbs (about 235,000 species)	Angiosperms which bear flowers and produce seeds from ovules that are fully enclosed within the carpel; after fertilisation, carpels may produce a fruit containing the seeds

Table 1.11 ***The main taxa of the Fungi***

Phylum	*Examples*	*Key features*
Ascomycota	yeasts, truffels, morels	develop sexually: spores formed inside a sac or ascus; asexual reproduction common
Deuteromycota	*Aspergillus*, *Penicillium*	no sexual life cycle known
Basidiomycota	mushrooms, toadstools, rusts	sexual reproduction; asexual reproduction rare; spores borne on club-shaped structures = basidia (sing. basidium)
Zygomycota	*Rhizopus* (black bread mould)	sexual and asexual development; multinucleate hyphae

Asexual reproduction produces genotypically identical progeny through **mitosis** and thus no variation (see Section 6.1). Variation does occur slowly through mutations, but most mutations are deleterious or lethal. The evolution of sexual reproduction involving a regular alternation between syngamy (the fusion of male and female gametes) and **meiosis** (the cell division that results in the segregation of genetic material required for gamete formation) led to much more diverse and rapid genotypic variation. The consequent mixing of genetic material on a regular basis formed the

Table 1.12 ***The main taxa of the Animalia***

Phylum	*Examples*	*Key features*
Porifera	sponges (about 5000 species)	mainly marine, asymmetrical animals lacking tissue layer organisation; feeding by special cells called choanocytes lining the saclike interior cavity
Cnidaria	jellyfish, sea anemones and corals (about 10,000 species)	two-layered simple body; may be solitary or colonial; gelatinous radially symmetrical bodies; feed by stinging prey
Ctenophora	comb jellies (about 100 species)	globular two-layered animals similar to cnidarians; gelatinous body using eight bands of cilia for locomotion
Platyhelminthes	flatworms, tapeworms and liverflukes (about 15,000 species)	three-layered solid, unsegmented bodies; bilaterally symmetrical worms; no anus; many important parasitic and disease-causing groups
Nematoda	roundworms such as *Ascaris*, hookworms	unsegmented, pseudocoelomates with mouth and anus. No cilia; important in all habitats and include some important parasites
Bryozoa	sea mats, sea mosses (about 4000 species)	collection of microscopic colonial groups also called Ectoprocta; feed using a lophophore (ciliated tentacles); individuals contained in a hard exoskeleton
Rotifera	wheel animals (about 2000 species)	small but ecologically important aquatic pseudocoelomates feeding with a ciliated crown which looks like a wheel; dominantly freshwater animals
Annelida	earthworms, ragworms, leeches (about 12,000 species)	serially segmented, coelomate worms; bilaterally symmetrical with mouth and anus; burrowing forms important in maintaining soils and sediments; leeches are often ectoparasitic
Arthropoda	crabs, spiders, scorpions, beetles, flies (about 1 million known species)	most successful of all animal phyla; bilaterally symmetrical with chitinous exoskeleton covering segmented bodies with paired, jointed walking appendages; many insects have wings; economically very important for diverse roles
Molluscs	snails, slugs, mussels, and squid (about 110,000 species)	soft-bodied, unsegmented coelomates, many of which have calcareous shells; most have a rasping tongue called a radula; many are important vectors of diseases; about a third are terrestrial
Echinodermata	starfish, urchins, brittlestars, sea cucumbers and crinoids (about 6000 species)	adults are radially symmetrical coelomates, often pentaradial; bilaterally symmetrical free-swimming larval stages; unique water vascular system operating tube feet and remarkable powers of regeneration
Chordata	fish, amphibia, reptiles, birds and mammals (about 42,000 species)	segmented coelomates with a notochord which becomes a dorsal nerve cord in the vertebrates; pharyngeal slits and a tail at some stage of life; about half of the chordates are terrestrial

basis for the variability of individuals, the raw material of evolution, and allows organisms to evolve rapidly in relation to environmental demands.

Life cycles became more diverse in eukaryotic forms and three main types of life cycles developed (Figure 1.10). In most protists, the only diploid cell is the zygote, which immediately undergoes meiosis and thus the life cycle is almost entirely haploid forms (zygotic meiosis). In animals, the gametes are the only haploid cells (gametic meiosis) in the typical life cycle. In plants and fungi, there is a more regular alternation between a multicellular haploid phase and a multicellular diploid phase (alternation of generations). The latter produces spores (sporic meiosis) leading to the haploid phase, which then produces gametes.

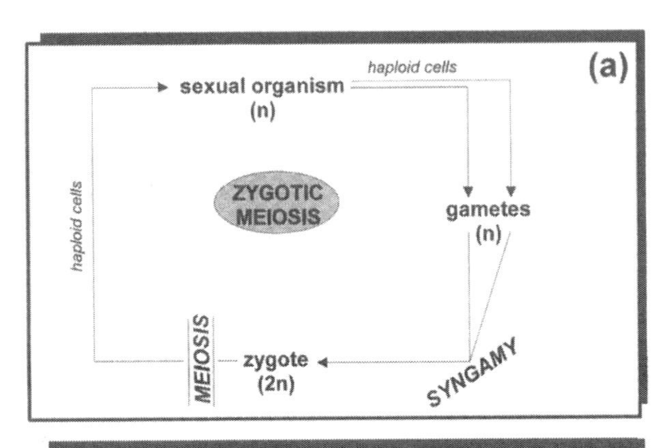

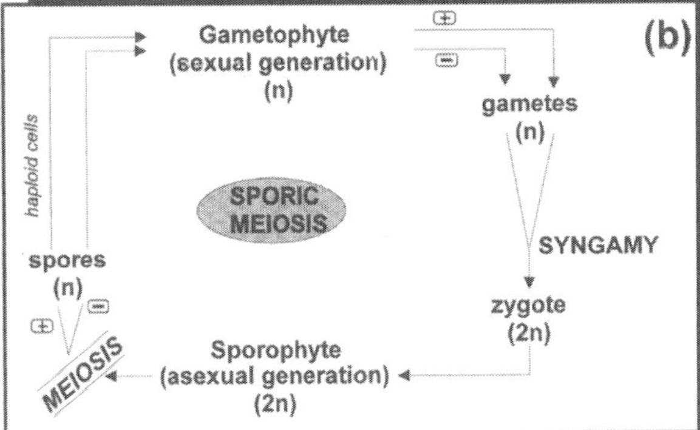

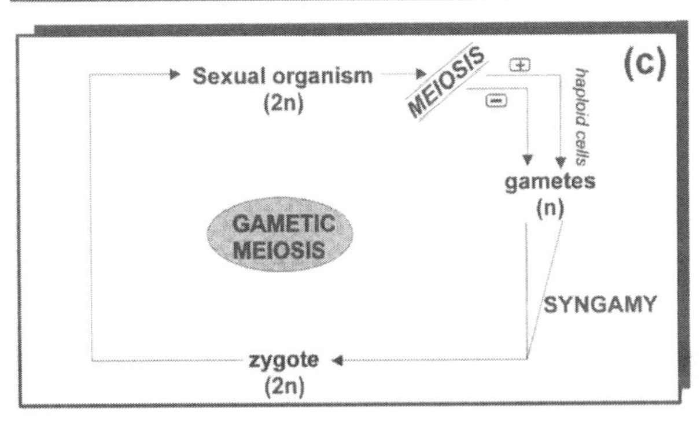

Figure 1.10 ***The three main types of eukaryotic life cycles. (a) Zygotic meiosis typical of algae in which the meiotic division occurs after fertilisation, resulting in a haploid sexual organism. (b) Sporic meiosis typical of plants with their alternation of generations and where the meiotic division occurs in the sporophyte generation. (c) Gametic meiosis, as in most animals, where the meiotic division occurs during the production of the gametes***

Summary points

- Living things combine various characteristics, most of which are not individually unique to living forms. The unique features of life are the ability to make and maintain complex structures from simpler ones (anabolism) and the use of hereditable materials in reproduction.
- Cells are the fundamental units of life and occur at two levels of complexity. Prokaryotic cells are relatively simpler than eukaryotic cells in both organisation and function. Multicellular organisms are mainly eukaryotes.
- The cell contains various structures which organise its activities. The plasma membrane forms the interface with the environment while the internal organelles of eukaryotes provide specialised units to make cell functions more efficient.
- Energy transformations form the engine of the cell and are governed by the First

and Second Laws of Thermodynamics. Cells acquire energy and use it to drive the cellular and higher activities vital for life.

- Cells are constructed from eleven main elements although others are important in trace quantities. The main molecules of life are water (with its unique properties) and the macromolecules grouped as carbohydrates, lipids, proteins and nucleic acids.
- Levels of organisation range from the individual through populations, communities and ecosystems. Defining an individual is not always simple and modular organisms are important in many ecosystems.
- The species concept is the basis for all classification (naming) of organisms. It is based upon the principle of reproductive isolation resulting in the formation of new species. The binomial classification system is hierarchical in structure.
- Six kingdoms of organisms are recognised currently: the Archaea, Bacteria, Protista, Plantae, Fungi and Animalia. The development of multicellularity was the key step in the evolution of the Plantae, Fungi and Animalia.

Discussion / Further study

1. We are all familiar with the forms of life that are of similar dimensions to ourselves and get a false impression of their relative abundance and significance. Devise a graphic representation of the relative global abundance of members of the different kingdoms to get a better conception of biological significance of these groups.
2. Try constructing a similar diagram for different ecosystems. Are there significant differences between them?
3. How many different levels of organisation are there within multicellular individuals? How have the evolution and sophistication of these levels of organisation facilitated the increase in complexity of the organisms interaction with the environment?
4. Make a diagram of the life cycle of (a) a seaweed, (b) a tapeworm parasite, (c) a fern and (d) a marsupial. Identify the key stages of the life cycle in terms of their reproductive stages, as in Figure 1.10.
5. Write out the taxonomic classifications of five different organisms as fully as possible. Look them up in textbooks but make sure that you understand how the system works.

Further reading

'The birth of complex cells'. C. de Duve. April 1996. *Scientific American* 274 (4), 38–45.
An excellent overview of the endosymbiotic theory of the origin of eukaryotes.

Biochemistry, 2nd edition. C.K. Matthews and K.E. van Holde. 1996. Benjamin / Cummings, Menlo Park, CA.
Clear explanations and illustrations of key biochemical processes and concepts. A fairly high-level text, not really for beginners.

The World of the Cell, 3rd edition. W.M. Becker, J.B. Reece, and M.F. Poenie. 1996. Benjamin/Cummings, Menlo Park, CA.
A very readable, student-orientated text on the biology of cells.

'Bacteria rule OK?' R. Lewin 3 June 1995. *New Scientist* 146 (1980), 34–38.
Genetic evidence for the endosymbiotic origins of eukaryotes from bacteria.

Five Kingdoms: An illustrated guide to the phyla of life on Earth, 2nd edition. L. Margulis and K.V. Schwartz. 1988. W.H. Freeman, San Francisco.
A well-illustrated overview of the five kingdoms recognised at that time.

Diversity of Organisms. C.M. Pond 1990. Hodder and Stoughton for the Open University.
A useful student-level guide to classification and diversity of life.

'Life unlimited.' R. Holmes 10 February 1996. *New Scientist* 149 (2016), 26–29.
An interesting and up-to-date overview of the problem of defining and identifying species of bacteria.

References

Mayer, E., 1970. *Populations, Species and Evolution: An abridgement of animal species and evolution*. Belknap Press of Harvard University Press, Cambridge, MA.

Raven, P.H. and Johnson, G.B. 1992. *Biology*, 3rd edition. Mosby-Year Book, Inc., Missouri.

2 The biosphere

Key concepts

- **The biosphere is structured by the interaction of organisms with the hydrosphere, lithosphere and atmosphere.**
- **The world's terrestrial habitats can be subdivided into biomes characterised by distinctive vegetation types. The concept has now been extended to the aquatic environment.**
- **The character of terrestrial biomes is determined mainly by temperature, rainfall, topography, and soil type.**
- **The character of aquatic biomes is based upon the source and movement of the water, its composition, the substrate type and the depth of the water.**
- **Biomes are useful classification devices but do not necessarily represent true communities or ecosystems, as outlined in Chapter 7.**

The *biosphere* (or ecosphere) is the planetary life support system. It is a 16-kilometre thick layer of living biological material with the most organisms found within only a thin (one kilometre thick) layer distributed over the earth's surface. E.O. Wilson (1992: 33) beautifully describes the tenuous nature of the terrestrial living layer thus:

> . . . imagine yourself on a journey upwards from the centre of the earth, taken at the pace of a leisurely walk. For the first twelve weeks you travel through furnace-hot rock and magma devoid of life. Three minutes to the surface, five hundred meters to go, you encounter the first organisms, bacteria feeding on nutrients that have filtered into the deep water-bearing strata. You breach the surface and for ten seconds glimpse a dazzling burst of life, tens of thousands of species of micro-organisms, plants, and animals within horizontal line of sight. Half a minute later, almost all are gone. Two hours later only the faintest traces remain, consisting largely of people in airliners who are filled in turn with colon bacteria.

Later chapters will discuss the vital processes operating within the biosphere. However, before doing so, it is worth while to attempt to classify the habitats in which life occurs. The physical environment may be divided simply into three major components:

- **Hydrosphere**: the water environments of the Earth (liquid and frozen, fresh and saline)
- **Lithosphere**: the soil and rock of the Earth's crust
- **Atmosphere**: the gaseous envelope surrounding the Earth.

The biosphere is structured into ecosystems by the interaction of organisms (the biota) with these physical divisions and the result is the obvious fact that organisms are not uniformly distributed within the biosphere. In fact, most organisms are absent from most environments most of the time. Why? Fundamentally, organisms are restricted to areas where the habitats (local environments) and potential lifestyles fit their particular evolutionary adaptations. The greater the physical differences between these areas, the larger will be the differences between their biotas.

Any system can be defined as a collection of interdependent components (subsystems) within a definable boundary. An **ecosystem** is a community of interdependent organisms together with the environment they inhabit and within which they interact but which is distinct from adjacent communities and environments: thus an oakwood is distinct from the farmland surrounding it. Once the living community and its environment become very large, however, they are best considered as *biomes*. The interactions of temperature, rainfall, soil type (especially nutrient availability) and local and geographical topography result in large areas with distinctive vegetation types which ecologists used as the basis for a classification of terrestrial habitats. These are biomes and are examples of very large ecosystems (see Chapter 7). Each is characterised by its own distinctive combination of flora, fauna and microbial communities, but is typically classified by the dominant form/s of vegetation, e.g. grassland or tropical rainforest. Although there is no definitive classification of biomes because they are not natural units, they are valuable as simple shorthand ecological descriptions of the world's major collections of biota. The biome concept has now been extended to aquatic habitats also.

2.1 The terrestrial biomes

Table 2.1 summarises the key features of the world's biomes in terms of their relative size and contribution to primary production (see Section 3.1). The most important factors determining the distribution of the major terrestrial biomes are temperature and rainfall: generally, the higher the rainfall and the temperature of an area, the greater the number and the larger the size of the plants it can support. At high latitudes, temperature dominates biome formation because of the restricted growing season, while in temperate regions rainfall becomes equally as important as temperature. Temperature varies relatively little in low latitudes and the amount and seasonality of rainfall largely determine the biomes. Altitude is a very important factor also, the changes in vegetation with increasing altitude being equivalent to moving from warmer to colder climates.

This section briefly considers the main features of each of the main terrestrial biomes; the classification used is in no way definitive and represents a compromise between complexity and completeness.

2.1.1 Tundra, the cold, boggy plains of the north

This is the most continuous of the Earth's biomes, forming a band around the North Pole interrupted only briefly by the North Atlantic and Bering Sea. Most, therefore,

Table 2.1 ***Relative primary productivity of the major ecosystems***

Ecosystem	*Area (millions km²)*	*Typical net primary production rate (g dry matter/m²/yr)*	*Estimated world net primary production (10⁹ dry tonnes/yr)*
Continental ecosystems			
Tropical rainforest	17.0	2200	37.4
Tropical seasonal forest	7.5	1600	12.0
Temperate evergreen forest (taiga)	5.0	1300	6.5
Temperate deciduous forest	7.0	1200	8.4
Boreal forest	12.0	800	9.6
Woodland and shrub land	8.5	700	6.0
Savannah	15.0	900	13.5
Temperate grassland	9.0	600	5.4
Tundra	8.0	140	1.1
Desert/semidesert shrub	18.0	90	1.6
Extreme desert, rock, sand and ice	24.0	3	0.07
Cultivated land	14.0	650	9.1
TOTAL TERRESTRIAL	*145.0*	*742*	*110.5*
Aquatic ecosystems			
Swamp and marsh	2.0	2000	4.0
Lake and stream	2.0	250	0.5
Open ocean	332.0	125	41.5
Upwelling zones	0.4	500	0.2
Continental shelf	26.6	360	9.6
Algal beds and reefs	0.6	2500	1.6
Estuaries and brackish waters	1.4	1500	2.1
TOTAL MARINE	*365.0*	*163*	*59.5*
Total Biosphere	*510*	*333*	*170*

Source: Adapted from Chiras 1994.

falls within the Arctic Circle although there are also small areas in the southern hemisphere and in high mountainous regions. The average temperature is –10°C and the region is characterised by a permanently frozen subsoil, the permafrost, and a seasonal melt of snow. It is a region of long, harsh winters and very short summers with long daylengths. The limited precipitation (only about 10 to 25 cm per year), together with low temperatures, flat topography and the permafrost layer, produces a landscape of broad, shallow lakes, slow-moving streams and bogs. Mosses, lichens, grasses and grass-like sedges dominate the tundra vegetation and are present in great abundance. Trees and other tall perennial plants are generally absent although there are numerous small perennial herbs and a few dwarf trees. The tundra teems with animal life at certain times of the year although the number of species present is relatively small, typical of a stressed habitat. Mammals such as reindeer, caribou, foxes, hares and lemmings are common and many species of wetland birds nest here in summer but migrate south in winter to avoid the winter extremes. Insects such as mosquitoes, blackflies and midges appear in vast numbers in summer.

2.1.2 Taiga, the evergreen boreal forests

Like the tundra, these are regions dotted by lakes, ponds and bogs and characterised by very cold winters, as in large parts of Scandinavia, Russia and Canada. However, the taiga has slightly more precipitation (about 50 cm per year) and a longer and warmer summer than the tundra, during which season the subsoil thaws and vegetation grows rapidly. The number of species present is larger than on the tundra but is still rather limited compared to lower latitudes. Drought-resistant evergreen conifers with needle-like leaves such as spruces and firs dominate, although some deciduous trees are common, e.g. aspen and birch. This boreal forest yields vast quantities of lumber for human activities. The soils are acidic, lacking in minerals and characterised by a deep layer of partially decomposed spruce and fir needles at the surface (see Box 1.2). The main herbivores are insects and many migrant birds are abundant in summer. Important mammals are moose, bears, wolves, lynx, martens and porcupines but smaller, herbivorous mammals such as rodents and rabbits are the most abundant forms.

2.1.3 Temperate deciduous forests

In temperate (mid-latitude) regions of all continents, precipitation varies greatly with location, continental interiors being drier and coastal areas being wetter. Hot summers and pronounced winters, together with rainfall of between 75 and 125 cm per year, produce deciduous forests with a rich topsoil and a deep, clay-rich lower layer. Broad-leaved hardwood trees such as oak and beech dominate although much of the original temperate deciduous forest world-wide has been removed by logging and land clearing. Where allowed to regenerate, it is now in a semi-natural state only. These forests originally contained a variety of large mammals such as wolves, bears and deer together with large numbers of small mammals and birds. Insects are abundant.

2.1.4 Temperate rainforest

Where annual rainfall is high (200–380 cm per year), often augmented by frequent coastal fogs, and winters are mild but summers cool because of proximity to the sea, a coniferous, temperate rainforest may develop. These relatively unusual conditions occur on the north-west coast of North America, south-eastern Australia and southern South America. The soils are rather nutrient-poor although organic content is often high. Large evergreen trees such as spruces and firs dominate these forests and are rich in epiphytic (supported on other plants) mosses, lichens and ferns growing non-parasitically on larger trees. It forms one of the most complex ecosystems on earth but is also one of the richest wood-producing biomes in the world and is, therefore, subject to considerable exploitation.

2.1.5 Temperate grasslands

Very large areas of the temperate regions are covered by grasslands characteristic of habitats where summers are hot, winters cold, and rainfall is relatively low and variable, averaging 25 to 75 cm per year. They are variously known as prairies (North America), steppe (Asia), pampas (South America) and veldt (South Africa). The soils contain much organic matter but if rainfall is very low, some minerals may accumulate just below the topsoil instead of being washed out, resulting in saline soils. Normal development results in the world's deepest soils and thus best agricultural land. Typically there are few trees except adjacent to watercourses and grasses grow in great profusion. They are natural ranges for grazing animals and contain large numbers of herbivores such as bison, cattle and antelopes, together with abundant insects. Burrowing animals are common. Overgrazing, however, can convert them to scrubland and poor farming practice can lead to loss of the topsoil. Combined with drought and wind, dustbowl conditions result (Box 2.1). Such land requires great care if it is to be exploited.

2.1.6 Tropical grasslands – the savannah

Grasslands, typically with widely scattered trees, are found in areas of low or seasonal rainfall with prolonged dry periods such as parts of Africa, South America and northern Australia. Temperatures vary little and seasons are regulated by rainfall rather than temperature, unlike temperate grasslands. Rainfall is between 85 and 150 cm per year, high but periodic. The soils are nutrient-poor usually because the parent rock is infertile. Savannah is characterised by extensive grasslands containing occasional thorny trees such as *Acacia*, whose thorns protect it from the abundant herbivores. The grasses and trees are fire-adapted to survive the frequent fires that occur naturally. The herbivore populations are massive, consisting of animals such as wildebeest, antelopes, giraffes, zebras and elephants, together with populations of carnivorous predators such as lions, cheetah and hyenas. Severe overgrazing converts it to desert.

2.1.7 Chaparral: shrubs and small trees

Wet, mild winters with very dry summers (e.g. the Mediterranean climate) in mid-latitudes such as in California, western Australia, parts of Chile and South Africa result in a vegetation dominated by a dense growth of evergreen shrubs that may also

Box 2.1

Desertification and land (mis)use

The conversion of range land, cropland and pasture into desert is desertification and it results in varying (moderate to severe) degrees of loss of productivity. Diverse factors contribute to desertification which usually occurs in arid, semiarid and dry–subhumid lands with distinctly seasonal climates. An obvious factor is naturally occurring drought but it may also result from anthropogenic activity such as overcropping, overgrazing and deforestation. It afflicts many countries, including the United States of America, Africa, Australia, Brazil, China and India and about 6 million hectares are estimated to become barren desert annually.

A prolonged drought from 1926 to 1934 in the windy and dry Great Plains region of the USA, together with land practices which resulted in the fertile topsoil being only loosely held, resulted in severe wind erosion and crop failure. In the 1930s, huge dust clouds resulted in large-scale animal deaths through choking and in 1934 a cloud of topsoil blanketed the eastern regions as far as 2400 km away. This area became known as the Dust Bowl and the result was the destruction of the agricultural economy of the region. Only through intensive conservation measures during the post-war years was the soil rebuilt and agriculture restored. Today, however, mini dust bowls are again being created in southern California and Texas due to overexploitation of the land.

Desertification is also taking place on a massive scale in parts of Africa. This is particularly true of the Sahel region, where a long-term drought has combined with overpopulation, overgrazing and poor land management to cause the southward extension of the Sahara desert. Estimates suggest that about 100,000 hectares are lost to desertification annually in Africa alone.

Table B2.1 ***Estimates of desertification***

	Susceptible dryland area	*Light and moderate desertification*	*Strong and extreme desertification*	*Total desertified*	*Percentage desertified*
Africa	1286.0	245.3	74.0	319.3	24.8
Asia	1671.8	326.7	43.7	370.4	22.1
Australasia	663.3	86.0	1.6	87.6	13.2
Europe	299.7	94.6	4.9	99.5	33.2
North America	732.4	72.2	7.1	79.3	10.8
South America	516.0	72.8	6.3	79.1	15.3
Total	5169.2	897.6	137.6	1035.2	20.0

Note: Data are in millions of hectares derived from UNEP 1992.

Source: Modified from Thomas 1995.

contain drought-resistant pine or scrub-oak trees. This is termed 'chaparral' in North America and has a thin, poor soil: fires occur naturally and frequently. During the rainy season, the area is lush and green but during the hot, dry summer the plants lie dormant and most have hard, small, leathery (*sclerophyllous*) leaves resistant to water loss. They are fire-adapted, the roots surviving the fires, and frequently grow best after the mineral release associated with the burning of the aerial parts.

2.1.8 Deserts

Deserts are very dry (arid) areas found in both temperate and tropical regions, mainly between 15 and 40 degrees north and south of the equator. They include the great deserts such as the Nabib and Kalahari deserts of southern Africa, the Sahara desert in northern Africa, the Arabian desert and the coastal Atacama desert of Chile and Peru. The low humidity of the desert atmosphere leads to large daily temperature ranges. Deserts vary greatly in the amount of rainfall received, generally less than 25 cm per year, and the timing of it is also unpredictable. Desert soil is low in organic content but has a high mineral content. Plant cover is very sparse at best and the soil is exposed to solar radiation and wind. Both annual and perennial species occur but perennials such as cacti, yuccas and Joshua trees are typical. Such species often have very small leaves or none at all, to minimise water loss while others may survive as underground bulbs or corms, having only a brief growing period following heavy rain. Succulents such as cacti in America and *Euphorbia* in Africa have thick cuticles, a low surface area:volume ratio and sunken *stomata*. These are the small, pore-like openings on leaves through which gaseous exchange takes place; they are open only at night to reduce water loss. Effective protection such as spines, thorns and toxic substances is common as a defence against grazing in this heavily stressed environment.

Desert animals have a formidable array of constraints to overcome: water and food are scarce, temperatures vary dramatically, sand makes locomotion and burrowing difficult and sand movement may cause burial. A diverse array of solutions to these problems has evolved, both behavioural and physiological. Many animals are small, remaining under cover or underground during the hottest parts of the day and hunting and foraging at night. Animals such as kangaroo rats survive entirely on water contained in foodstuffs and released during metabolism (*metabolic water*). The biomass of life is low and the biota highly specialised.

2.1.9 Tropical rainforests

These occur near the equator where rainfall is high throughout the year and temperatures are consistently warm: in South and Central America, West and Equatorial Africa, South-east Asia, Indonesia and north-east Australia. The annual rainfall is between 200 and 450 cm, much of it derived from transpiration (loss of water vapour by plants) from the lush vegetation itself. Tropical rainforests are considered to contain the greatest diversity of life of any of the world's biomes yet commonly occur in areas with old, mineral-poor soil extensively leached by the rainfall. The result is a poor soil with low nutrient and organic contents.

Figure 2.1 ***The layered structure of the rainforest canopy***

The forest structure is highly layered into canopies (Figure 2.1). The upper (40–50 m high) and middle (30–40 m high) canopy layers form a continuous screen of leaves, effectively preventing most of the light from reaching the floor. This produces a dark, warm and very moist microclimate beneath the canopy in which decomposition processes are rapid but vegetation is relatively scarce. The rapid decomposition of plant litter is very important in the nutrient cycling of these forests. Despite the poor soils, tropical rainforests are very productive (see Table 2.1) and highly diverse, no one species of plant or animal dominating. One of their main ecological characteristics is their fragility; since once destroyed, they are very slow to regenerate, if it happens at all.

2.1.10 Tropical seasonal forests

These occur where a distinct dry season occurs within an otherwise humid tropical climate. The trees may lose their leaves during this dry season and they are markedly less diverse than tropical rainforests. They usually occur where climates include monsoons (seasonal rainfalls), such as India, South-east Asia, East and West Africa, Central and South America, the Caribbean and northern Australia. As the climate gets drier and soils get poorer, such as in northern South America, the Caribbean and South Africa, tropical broad-leaved woodlands containing small trees replace them. The canopy is only from 3 to 10 m high and consists of trees and shrubs with thick, fire-adapted bark and twisted branches.

2.2 The aquatic biomes

The terrestrial biomes described above are influenced largely by temperature and rainfall patterns together with soil type and wind. Aquatic systems also have their major determinants which are:

- the amounts of dissolved materials present in the water (the solutes)
- the depth of water
- the availability and quality of light
- the nature of the bottom substrate (e.g. rock, sand or mud)
- water temperature and circulation patterns.

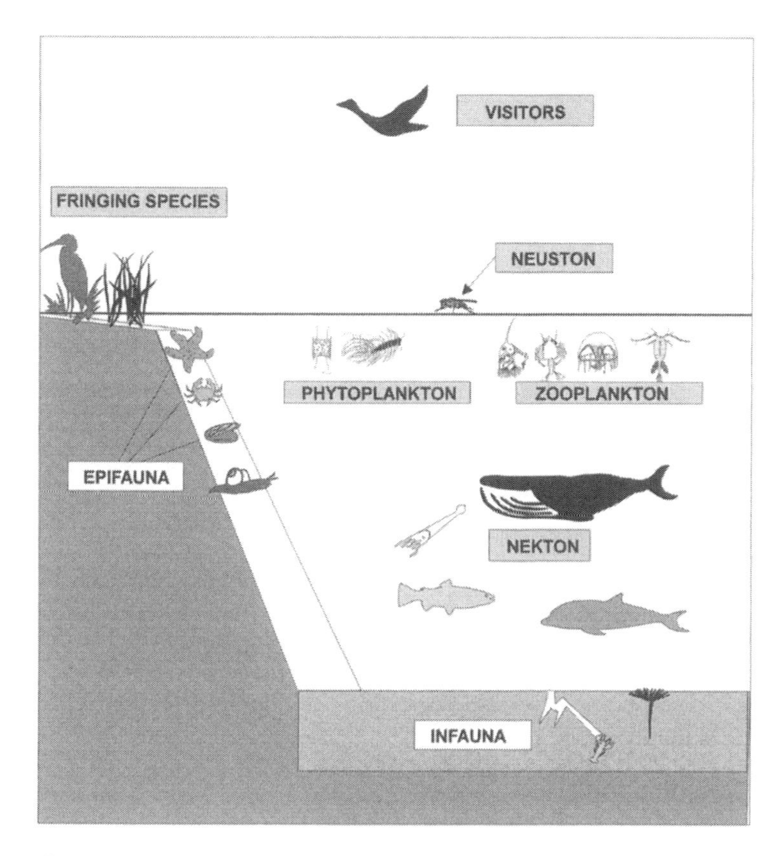

Figure 2.2 ***A basic classification of marine ecological groups***

The most fundamental division in the aquatic realm is that between freshwater and saltwater environments, with estuaries and brackish areas standing as intermediates (see Chapter 4 for details). The **salinity** (the concentration of dissolved salts in the water – see Table 4.1) exerts a major influence on the biota. Fully saline sea water is very similar, both osmotically and ionically, to the body fluids of most organisms and is much less physiologically demanding than is fresh water, which contains very little dissolved material. Fresh water, therefore, causes particular physiological difficulties and only species adapted to these problems can inhabit freshwater environments.

Water greatly interferes with the penetration of light, particularly the photosynthetic wavelengths (see Figure 4.8), so photosynthesising forms must remain near to the surface to function. The nature of the substrate determines what can live in or on it. Water temperature and circulation are major determinants of the physico-chemical characteristics of the water body and on the distribution and survival of its biota.

Aquatic habitats contain three main ecological groups of organisms (Figure 2.2):

1 **Plankton**: organisms incapable of swimming from current system to current system. They are effectively free-floating species although many have limited powers of locomotion, usually used to maintain their vertical position in the water column. They are usually small or microscopic organisms, subdivided according to size, although some jellyfish may reach several metres in diameter and many metres in length. They are generally further subdivided into two major categories:
   - **phytoplankton** (plant plankton), photosynthetic cyanobacteria and free-floating algae (mainly diatoms) which are the main primary producers in most aquatic systems and
   - **zooplankton** (animal plankton), the non-photosynthetic species that include the **holoplankton**, species which spend their entire life cycles in the plankton and **meroplankton**, larval forms of species which are not planktonic as adults.

2 **Nekton** are the stronger-swimming species that are capable of swimming between current systems and include fish, squid, turtles and whales. They are a heavily exploited component of the system.

3 **Benthos**, the plants and animals that live on the bottom (**epibiota** including seaweeds, barnacles and lobsters) or burrow into the substrate (**infauna** such as worms and small crustaceans).

2.2.1 Wetlands

There is a large variety of wetland systems around the world and they are quite common components of tundra and taiga. They are areas where water is more significant than climate in the determination of the biota; this leads to considerable similarities in different regions. Although the species in the communities are often different, the basic structures of the system are very similar. These systems are difficult to classify since climate and geographical location are not as significant as for terrestrial biomes. Chapman and Reiss (1992) divide them on the basis of the source and nature of the water maintaining the system and recognise four types:

1 The sea – sea water is highly saline and tidal

2 Streams and rivers – sediment-rich and seasonally flooded

3 Drainage from surface runoff or groundwater – usually nutrient-rich and alkaline

4 Direct from rain or snow – rainwater is nutrient-poor and acidic due to the presence of dissolved carbon dioxide and, in polluted areas, sulphur dioxide.

The following classification is based upon each of these categories of water.

2.2.1.1 Marine wetland systems

Two main marine systems are recognised.

1 **Mangrove swamps** fringe tropical and subtropical sheltered coastlines where tides inundate the swamp with marine or brackish water every high tide. The mangrove tree is well adapted to living in salty environments and about seventy species occur throughout the world.

2 **Salt marshes** are found on sheltered coastlines at higher latitudes where mangroves are excluded by being frost-intolerant. Inundated regularly by high tides, they consist of a patchwork of low vegetation separated by tidal creeks. Dominated by grasses and rushes, the soils are high in phosphates but low in nitrogen and mats of nitrogen-fixing cyanobacteria are important in the nutrient balance of many of these systems. Salt marshes are quite productive but are generally very sensitive to damage and disturbance.

2.2.1.2 Floodland ecosystems

Deriving their water from rivers, these wetlands are often very seasonal. They tend to occur in lowland flat-bottomed valleys containing large mature river systems. Flooding carries clays, silt and mineral nutrients into the systems and this frequent addition of new material makes flooded valley soils nutrient-rich. However, during the flooding, soils may become temporarily anaerobic, an important limiting factor that determines the communities living at the site. The natural vegetation in temperate floodplains is mixed deciduous forest, usually with a rich fauna. The frequency and duration of flooding events determines the community structure.

2.2.1.3 Swamp and marsh systems

These occur in areas with impeded drainage where water runs off the surrounding land and collects or where the water-table is close to the surface. These areas are flooded continuously except in exceptional, very dry years, and are very variable in size and form, depth of soil and plant community structure. Swamps have trees as the dominant form of vegetation while marshes have large open areas of grasses, sedges and reeds. Flooded peat marshes with non-acidic soils are known as fens in Britain and alkaline bogs in America

Swamp can only regenerate when it dries out, although this need only happen about once in a hundred years because of the longevity of the trees. Marshes are usually nutrient-rich with a fairly high pH and are common in temperate zones. Most leaf matter falls to the wet marsh surface and decomposes, the residue frequently forming a peaty layer.

2.2.1.4 Bog systems

Receiving water only from rainfall which has a low nutrient content, bogs are nutrient-poor, acidic systems with a pH of 3–4. The dominant species are mosses of the genus *Sphagnum*. Bogs arise mainly in the temperate and boreal regions. The bog may slowly rise to form a dome of peat called a *raised bog*, usually confined by some topographical feature. If rainfall is high (more than 100 cm per year), *blanket bog* can develop with a peat layer several metres thick. This type of bog is vulnerable to erosion once the surface is damaged by grazing, trampling or pollution. Bog communities are slow-growing with slow production of plant material (primary production) and very limited faunas.

2.2.2 Freshwater life zones

Freshwater ecosystems fall into two distinct categories: flowing-water environments (lotic) and standing-water environments (lentic).

2.2.2.1 Lotic systems

have a unidirectional flow of water but they vary greatly in their scale from the small mountain stream to the Amazon river. Characteristically, lotic systems have a *molar* (eroding) action on the surfaces over which they flow and this results in marked variations in the water's composition and its sediment load. The flow rate largely determines the sediment load because, as the flow slows, smaller and smaller particles fall out of suspension. Thus fast-flowing upland streams have bedrock and boulder substrates while the slow-moving mature rivers have sand and mud deposits. The combination of current speed and substrate type has a marked effect on the flora and fauna in the different regions of a lotic system, leading to diverse structural, behavioural and life-cycle adaptations. Table 2.2 contrasts the general features of lotic and lentic systems. An important feature of most lotic systems is their dependence on detrital material derived from outside the system (i.e. from adjacent land or from lentic systems) for most of their food. This is because planktonic and benthic primary producers are often relatively insignificant in this system.

Table 2.2 ***The key features of lotic (flowing) and lentic (standing-water) systems***

Lotic systems	*Lentic systems*
unidirectional water flow – currents are gravitationally induced	no unidirectional flow of water – currents are largely wind or temperature induced
migration of environments occurs over time due to erosion.	evolution of environments occurs due to infilling by sediments
brook > stream > river	lake > pond > swamp
system biology mainly dependent upon:	system biology mainly dependent upon
• substrate chemistry of watershed	• surface:volume ratio of water body
• speed of water flow	• nutrient status
• depth of water	• plankton during early stages
• nature of river bed	• benthos during later stages
migration of system due to:	infilling of system due to:
• molar action of suspended particulates	• wind-blown debris
• chemical erosion of substrate rocks	• sediment input from feeder lotic systems
	• wave erosion of shoreline
	• death of biota

2.2.2.2 Lentic systems

comprise the standing-water systems including lakes (lacustrine habitat) and ponds which, although they may have very slow-moving wind and convection-induced currents, never have unidirectional flows. The major determinants of lentic systems are their permanence, the nature of any stratification of the water body, and their nutrient status. If nutrients are abundant, lakes are termed eutrophic and when scarce, they are oligotrophic. Large lakes have three basic life zones, the littoral, the limnetic and the profundal; smaller and shallower water bodies lack the profundal zone.

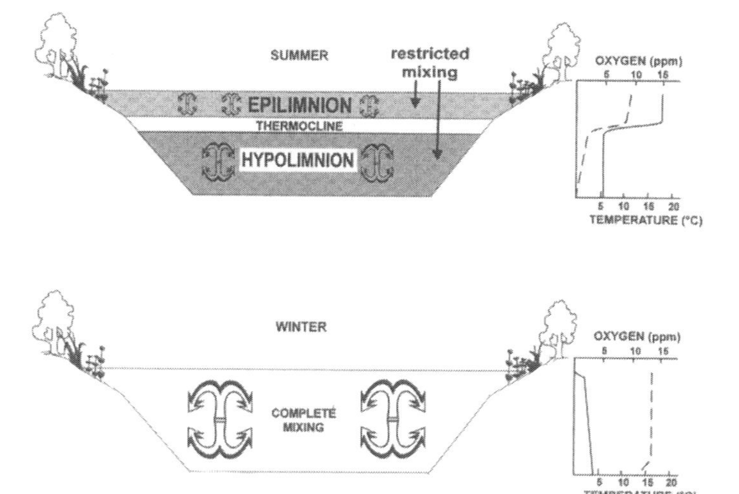

Figure 2.3 ***The changes in lentic freshwater systems resulting from thermal stratification typical of temperate regions. The thermocline is formed during the spring months and breaks down again during autumn months, exerting major effects on the nature of circulation in the system. This in turn affects the temperature (solid line) and oxygen levels (broken line) in the water column (insets)***

- The **littoral zone** is the shallow, wave-splashed zone along the edge of the system. Its biota includes a variety of vascular plants and algae and it is a highly productive zone because of nutrient runoff from adjacent soils. It also has a diverse fauna including numerous insect larvae.
- The **limnetic zone** is the open-water area away from the shore extending downwards to the limit of sunlight. It is populated mainly by the planktonic community and their predators. It is influenced greatly by the possible formation of a *thermocline* (Figure 2.3), a temperature-based density layer (a type of *pycnocline*) which can effectively divide the upper (epilimnion) and lower layers (hypolimnion) of a lake into separate water masses. This is particularly significant since it restricts nutrient distribution.
- The **profundal zone** is the dark and deep zone below the limnetic zone where most of the microbial decomposition processes vital to nutrient recycling take place. This zone tends to be both mineral-rich and oxygen deficient and contains relatively few macrofaunal species.

2.2.3 Marine life zones

Marine life zones are similar to those in lakes but the scale of the oceans is so much greater that further subdivision of marine zones is usual. Figure 2.4 shows the main generalised life zones.

As in lentic freshwater systems, the boundary region between land and water is termed the **littoral** (or intertidal) zone. However, this region is influenced not only by wave action but by the tides, mass movements of water caused by the gravitational effects of the sun and moon. Although nutrients, light and oxygen are plentiful, this is a highly stressed environment influenced by the variation in the ranges of heat, light and desiccation factors (Table 2.3). This habitat may consist of mud, sand or shingle but where waves or currents are strong, the shore usually comprises bedrock or boulders.

Table 2.3 ***The tidal environment and its key features***

Tide in	*Tide out*
Temperature relatively constant	Temperature fluctuation may be considerable and rapid
No desiccation problem	Desiccation danger considerable (including physiological desiccation in evaporating rock pools)
Gases such as oxygen and carbon dioxide common and available	Oxygen plentiful but may be relatively unavailable. Carbon dioxide levels very reduced
Light intensity reduced, allowing photosynthesis	Light intensity may be very high, producing photo-inhibition and heating effects
Food is plentiful and available	Feeding usually inhibited or prevented
Locomotion and feeding possible	Locomotion and feeding inhibited or prevented
Support for soft-bodied forms through buoyancy	No support so that weight may inhibit other activities
No osmotic or ionic problems	Osmotic and ionic problems common and may be severe e.g. when it rains

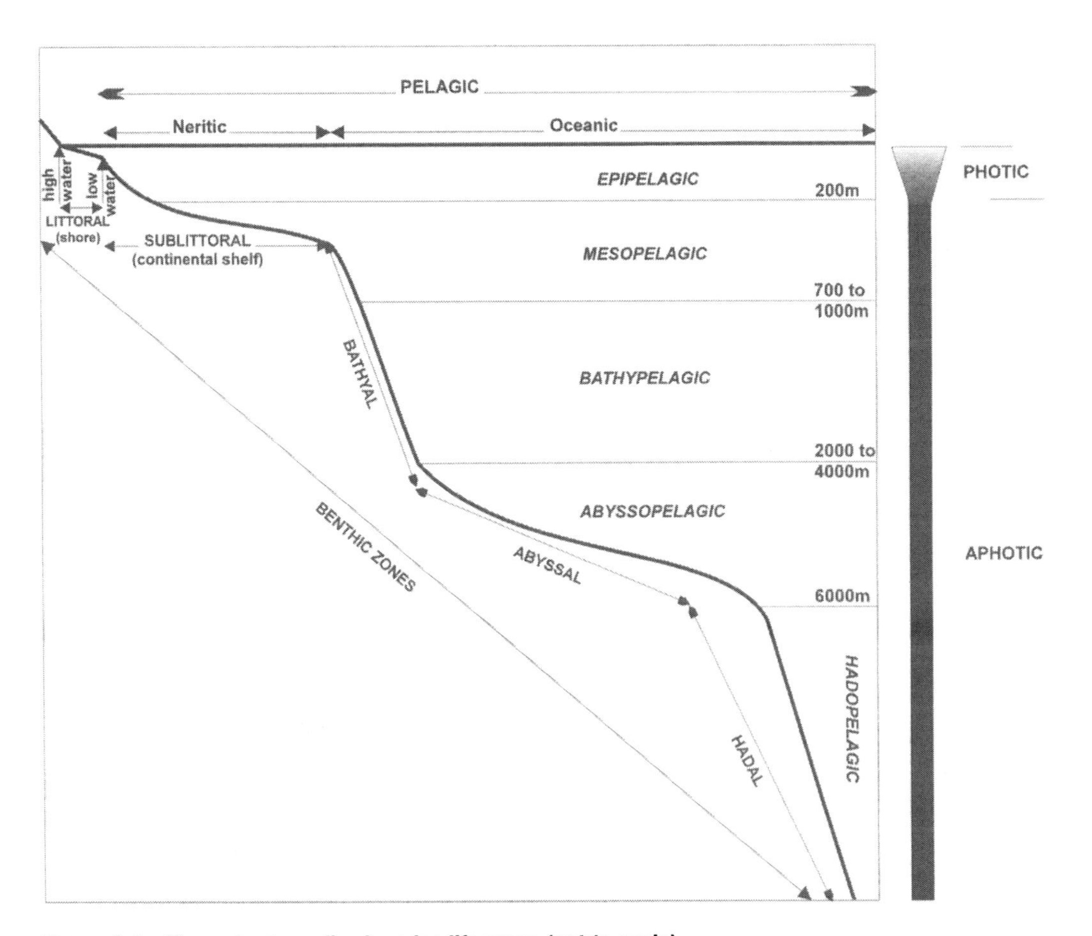

Figure 2.4 ***The main generalised marine life zones (not to scale)***

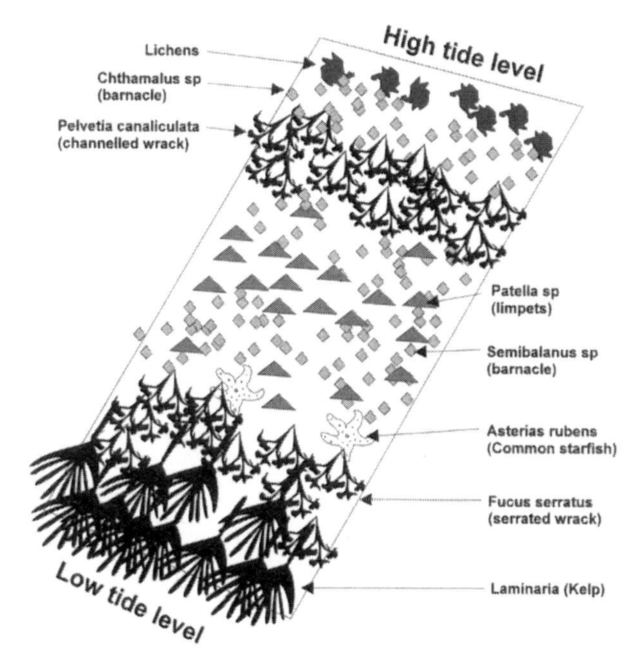

Figure 2.5 ***Diagrammatic representation of zonation on a typical semi-exposed rocky shore in Britain***

2.2.3.1 Marine rocky shore

This typically shows zonation of both mobile and sedentary organisms in response to tidal gradients of conditions (Figure 2.5). Living forms are on the surface (epiflora and epifauna) although irregularities in texture and the occurrence of stones and boulders, crevices, etc. provide frequent microhabitats with their own characteristics: one of the most familiar is the rockpool. Primary production by seaweeds is often considerable.

2.2.3.2 Marine sandy shore

Although also exhibiting zonation, the main difference from the rocky shore is that the mainly animal biota is not visible on the surface. Instead, it burrows into the sediment (infauna), where species also become stratified by depth. There are few plants, the majority being microscopic diatoms and some green algae that live between the sand grains (the interstitial spaces) near the surface. This biome is dependent for its food resources upon the planktonic production of the overlying water when the tide is in or on detrital material brought in from other habitats. The ability to burrow alleviates to some extent the heat, light and desiccation problems so important on rocky shores. However, living in the sediment has its own difficulties. These include abrasion, instability (hence the lack of macroalgae) and potential lack of nutrients, the sand being composed mainly of silicates and calcareous materials. The main factor determining the biota is the particle size of the grains. Most organisms prefer fine-grained sediments, coarse sediments usually containing only transient forms because of their inherent instability. The rich life often found burrowed in fine and medium-grade sand is important for the support of populations of wading birds and young fish.

2.2.3.3 Marine mud-flats

These are usually associated with freshwater inputs of sediment and are composed of fine silts and clays. These sediments tend to retain organic matter and, because of the tiny interstitial spaces and the relatively high organic content, they tend to have severely restricted circulation. This renders them anaerobic (lacking in oxygen) only a few millimetres beneath the sediment surface. Only comparatively few detritus-feeding species adapt to survive in these conditions but those that do are often present in very large densities because of the abundance of detrital organic matter. These areas are often very important as sources of food for birds.

The shore is the upper limit of the benthic environment, those parts of the system associated with the sea floor. Below the tidal zone are further benthic zones (see Figure 2.4) but these can be subdivided into the continental shelf benthos and the deep ocean benthos.

2.2.3.4 Benthos of the continental shelf

These are the organisms living down to about 200 m, including shallow waters with sufficient light penetration to allow photosynthetic activity by macroalgae. This is the zone of the kelp forests, large, highly productive aggregations of brown algae such as *Laminaria* and *Macrocystis*, the latter growing up to 50 metres in length at rates as high as 25 cm each day. Their primary production enters the marine system mainly as detritus, however, as relatively few animals graze them. The continental shelf benthos is a diverse biota comprising many phyla and is often very productive, being exploited significantly by humans for its diverse marine products.

2.2.3.5 Deep ocean benthos

Extending from the edge of the continental shelf to the deepest ocean trenches, these communities are almost entirely heterotrophic. They depend upon the rain of detrital material (marine snow) or food parcels (dead animals) falling from the sunlit layers above. The only primary production here occurs in the remarkable chemosynthetic communities found associated with volcanic vents (fumaroles). The environment is characterised by a real monotony of physico-chemical conditions and by the restricted food supply, having been likened to a desert in terms of faunal densities and biomass. All are cold (less than 4°C), dark and at high pressure because of the weight of the water above. Knowledge of the organisms of this region is sparse but evidence suggests that growth rates and population densities are much lower than those of shallow water organisms.

2.2.3.6 The pelagic environment

The open water habitat is termed the **pelagic** environment and is divided horizontally into two provinces: the **neritic** (coastal) one is bounded by the edge of the continental shelf (about 200 m depth) while the **oceanic** one comprises those regions of water beyond the continental shelf. The oceanic province occupies by far the majority of the world's oceans. Vertical subdivisions are also recognised. The uppermost well-lit layer where photosynthesis takes place is the euphotic zone and is where most of the ocean's primary productivity occurs. Below that layer, down to about 1000 m, is the twilight zone, so called because sunlight is present only in very small amounts: light intensity decreases with depth until it disappears totally at about 1000 m. The twilight zone contains a diverse array of animal forms, many of which have special adaptations to overcome the problems associated with low animal densities (privation). This includes bioluminescence, the production of light biochemically by many organisms (Box 2.2).

Below this 1000 m boundary, which is a biologically based division rather than one based solely on physical characteristics, further subdivisions use mainly topographical features (see Figure 2.6). Most of the scarce life that exists in the monotony of the deep ocean depends upon food materials derived, often through the food web, from the sunlit layers above. Animals in the abyssal depths are adapted to permanent darkness, very stable physical and chemical conditions and scarcity of food: many are predators or scavengers living in highly dispersed populations.

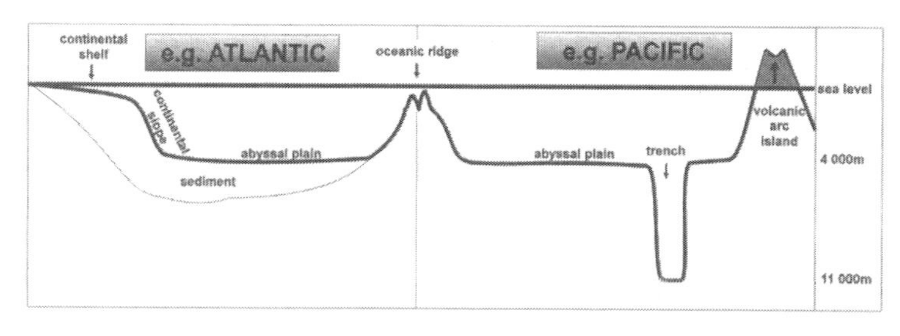

Figure 2.6 ***The main topographical features of the seabed of the world's oceans***

2.2.4 Estuarine and brackish environments

This is a special category of aquatic ecosystem consisting of shallow, partially enclosed areas where fresh water and sea water meet and mix. Brackish environments are usually relatively stable but diluted seawater systems, often forming lagoon-like areas adjacent to the sea. Estuaries are highly variable environments where conditions change rapidly both with space and time due to variations in the relative inputs of fresh water and sea water: these vary with rainfall and with tidal movements respectively. They also act as sediment traps because, when fine particles in suspension in fresh water meet sea water, electrostatic interactions cause **flocculation** (clumping) to occur, depositing mud. Thus, these systems tend to become filled in and may also accumulate many pollutants which adsorb to the fine sediment particles.

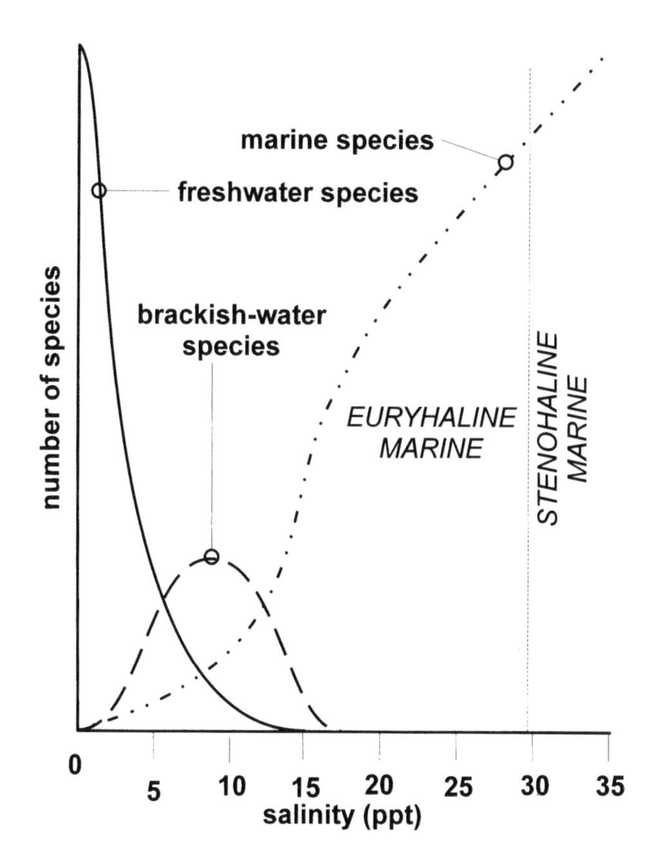

Figure 2.7 ***The effects of salinity variation on the nature and abundance of species***

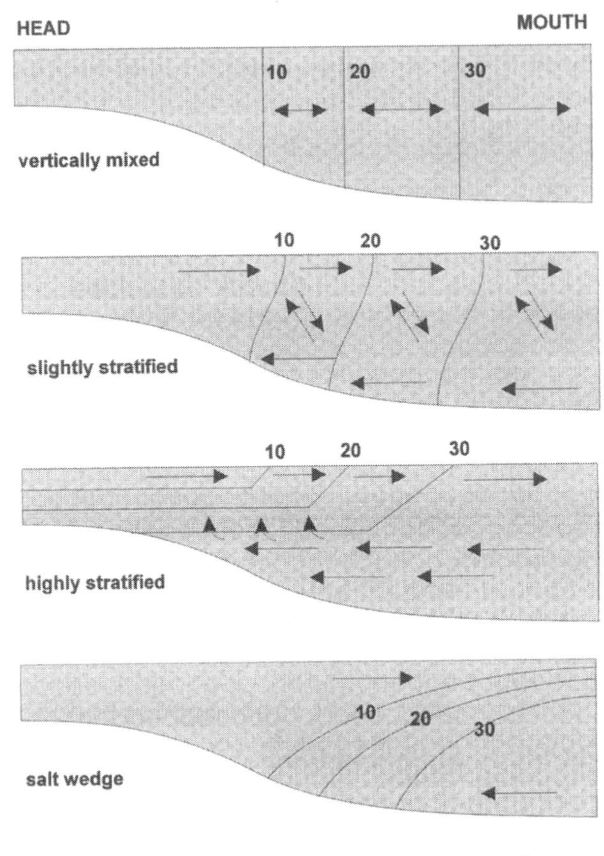

Figure 2.8 ***A summary of the main types of estuarine circulation. The values represent salinity isohalines and the arrows indicate the direction of current flow***

Diversity is reduced in these highly stressed environments (Figure 2.7) but the abundance of adapted species may be very high because they are usually nutrient-rich and detritus-rich environments. Productivity is usually high because of the abundant nutrients washed into them, enhanced by the shallowness that allows light to penetrate most of the water in the basin. Estuaries are often important as nursery sites for fish and crustaceans and as feeding areas for diverse bird populations.

One of the most significant features of an estuarine system is the nature of the mixing that takes place: Figure 2.8 illustrates the main categories of estuarine circulation. A fully mixed estuary such as the Tay Estuary in Scotland is much less sensitive to human inputs than a stratified estuary such as the Clyde Estuary. The topography of the system and the volumes of fresh water and sea water entering the system largely determine the type of mixing.

2.2.5 Coral reefs

Coral reefs are a special form of benthic community found in warm, shallow, nutrient-poor but clear, sea water in tropical and subtropical parts of the world. The corals forming the basis of this system are largely symbiotic, containing single-celled algae within their bodies which contribute to the nutrition of the corals by photosynthetic activity. For this reason, corals are able to flourish in the nutrient-poor waters typical of many tropical areas. Indeed, the introduction of nutrients by human activity (sewage) often leads to the destruction of the reef by encouraging seaweeds which would not otherwise be able to grow (Box 2.3).

The corals themselves provide the physical basis for a complex and dynamic system of other forms, making them the marine equivalent of tropical rainforest in terms of diversity. They are considered to be one of the most productive ecosystems on Earth.

Box 2.2

Biological light in the ocean depths

The most prevalent light-producing organisms in the oceans are the bacteria and dinoflagellates which utilise a special biochemical system termed the luciferin-luciferase system to produce bioluminescent light. The system probably evolved in pre-eukaryotic times as a mechanism to deal with the toxicity of the oxygen produced by the first photosynthetic organisms; the production of light was an accidental consequence. Today, luminous bacteria live at all depths of the oceans and some squid and many fishes have organs that contain symbiotic cultures of luminescent bacteria (*Photobacterium* spp.) as a means of producing light. Many organisms, however, have developed bioluminescent capabilities of their own. Although bioluminescence is relatively rare in terrestrial ecosystems, the vast majority of oceanic animals that inhabit the region between 500 and 1000 m deep appear able to produce light. They use a variety of mechanisms and fulfil diverse purposes. These include escape from predation, camouflage, shoaling and species identification, communication and hunting by light beams or lures. Bioluminescent light is present down to almost 4000 m but the most luminous depths are in the mesopelagic zone.

Many animals have complex light organs (photophores) of precise optical design and performance (Figure B2.1) while others extrude luminescent materials such as the luminous 'ink' in the deep-sea squid *Heteroteuthis*. Although bioluminescent light is blue-green with a wavelength of between 470 and 510 nm, many organisms can colour their light using special filter cells. Species-specific patterns of light organs are typical of the mesopelagic fish and squid species. Some voracious hunters such as *Malacosteus* (Figure B3.1, Box 3.3) produce a beam of red light from a light organ beneath their eyes and use this for hunting their prey. Since most organisms living in the sea cannot detect red light, this light beam is invisible to them. Some fish species use their ventral light organs for countershading by producing just enough light to eliminate any shadow produced when seen from below against the background of residual daylight.

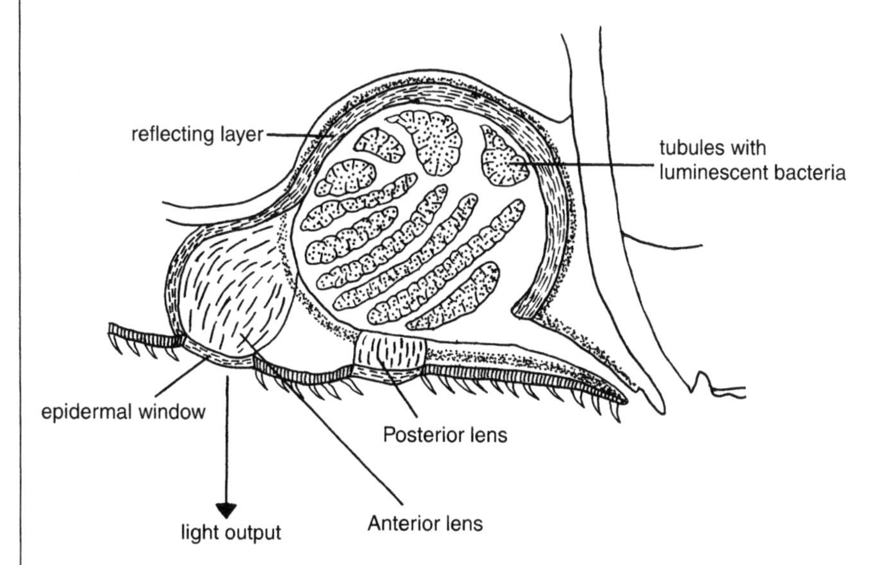

Figure B2.1 ***The structural elements of a bacterial light organ of* Malacocephalus laevis *(a deep-sea fish) (adapted from Marshall 1979)***

Box 2.3

Coral reefs – the ocean's rainforests

Coral reefs are highly complex and diverse ecosystems that develop in the warm, shallow and nutrient-poor waters typical of many tropical and subtropical oceans. They are highly productive systems and are often likened to rainforests in terms of both diversity and productivity. This very high biodiversity reflects the large variety of available microhabitats and niches within the complex physical structure of a reef. It has been estimated that at least a third of all marine fish are associated with coral reefs. Many reef dwellers are primary producers, either marine plants or symbiotic algae (*zooxanthellae*) living within or between the cells of the corals themselves.

Coral reefs grow slowly and are easily disrupted, thriving only in clear, warm (always above 18°C), nutrient-poor water with a constant high salinity. They comprise a calcareous framework made up mainly of the interlocked and encrusted exoskeletons of reef-building corals, calcareous red algae and other associated organisms. They occur in a variety of broadly distinct forms, namely fringing reefs, barrier reefs and atolls. Fringing reefs are continuous with the shoreline of the associated landmass although a shallow and narrow channel may develop behind the reef. A barrier reef lies some distance offshore from its associated landmass, separated from it by a lagoon generally more than 10 m in depth. An atoll is an offshore reef formation, roughly circular in shape and surrounding a central lagoon, but no exposed landmass is associated with it. The sustained existence of flourishing and productive coral reefs in nutrient-poor oceanic waters appears at first to be paradoxical. However, the symbiotic corals have developed highly effective internal mechanisms for the recycling of nutrients within the system and so require little external nutrient supply. However, despite their high diversity and apparent stability, they are very susceptible to disturbance and they are being destroyed or damaged in many parts of the world. This is largely a consequence of human activity although these are also natural problems such as hurricanes, predation by crown-of-thorns starfish (*Acanthaster* spp.) and ocean warming (which can produce bleaching). Many of the greatest threats come from the fact that socio-economic forces often encourage the immediate exploitation of reef resources, particularly in developing countries. Another major problem is the increase in sediments and nutrients in the water as a result of urbanisation, deforestation, agriculture and poor land management. Sediment smothers the reef and blocks out the vital sunlight, while the increase in nutrients leads to colonisation of the reef by macroalgae and again it becomes smothered by this new growth.

Some 300 coral reefs in 65 countries are protected as reserves or marine parks but protecting, managing and restoring reefs is very difficult and expensive. Ten per cent of the world's reef area is estimated to have been degraded beyond recovery already and about another 30 per cent is under severe threat. These are very imperfectly understood ecosystems so prediction and management are particularly difficult.

Thus, the diversity of the major habitats has been described and some of the key environmental characteristics discussed briefly. The following chapters will now look further into the functioning of such systems, starting with the basic principles associated with the movements of energy and materials and then examining the biological components from the viewpoint of increasing levels of complexity.

Summary points

- The biosphere is structured by the interaction of organisms with the hydrosphere, lithosphere and atmosphere. Organisms are restricted to areas where the habitats fit their evolutionary adaptations.
- The world's major habitats can usefully be subdivided into biomes as described here. There is no definitive classification of biomes because they are not natural units but descriptive, ecological shorthands for collections of biota.
- Terrestrial biomes are determined mainly by temperature ranges and patterns and quantity of rainfall. At high latitudes, temperature is usually the most significant feature while in the tropics, rainfall is usually the key factor. In temperate regions, they are both equally significant.
- Aquatic biomes are based upon the source and nature of the water in the system. Key factors are the depth and composition of the water, light availability, the nature of the substrate, water temperature and circulation patterns.

Discussion / Further study

1. Consider the impact of urban and industrial sewage pollution on (a) a freshwater stream, (b) a small lake system and (c) an estuary. How can the biotas of these systems cope with the input of organic materials and other pollutants?
2. What are the key features of a tropical rainforest? Why is slash and burn agriculture so damaging to such a system?
3. Make a scale diagram to show the relative scale of the biosphere in terms of the Earth's physical dimensions.
4. Identify the special structural, behavioural and physiological adaptations of two plants and two animals living in a desert.
5. Why did the building of the trans-Alaska oil pipeline pose particular environmental problems?
6. Contrast the problems of living in terrestrial and aquatic particulate sediments.

Further reading

Light in the ocean's midwaters'. B.H. Robinson. July 1995. *Scientific American* 273 (1), 50–56.
An up-to-date review of the occurrence of bioluminescence in the deep sea.

The Living Planet: A portrait of the Earth. D. Attenborough. 1984. William Collins and Sons/BBC.
An excellent illustrated account of the biomes.

Ecology and Field Biology, 5th edition. R.L. Smith. 1996. HarperCollins College Publishers, New York.
A good ecological text dealing with all aspects of ecology in reasonable detail.

Fundamentals of Aquatic Ecology. R.S.K. Barnes, and K.H. Mann. 1991. Blackwell Scientific Publications, Oxford.
Excellent scientific overview of the aquatic biomes and their characteristics.

Introductory Oceanography. H.V. Thurman. 1994. Macmillan Publishing, New York.
Excellent introduction to the marine environment.

Ecology of Fresh Waters: Man and medium. B. Moss. 1988. Blackwell Scientific Publications, Oxford.
Very readable book suitable for various levels of interest.

'Desertification'. J. Wellens and A.C. Millington. 1992. In *Environmental Issues in the 1990's*, ed A.M. Mannion and S.R. Bowlby. John Wiley and Sons, Chichester.
An excellent review of the problem.

Global Biodiversity: Status of the Earth's living resources. World Conservation Monitoring Centre. 1992. Chapman and Hall, London.
A compendium of facts and information on this topic.

References

Chapman, J.J. and Reiss, M.J. 1992. *Ecology: Principles and applications*. Cambridge University Press, Cambridge.

Chiras, D.D. 1994. *Environmental Science: Action for a sustainable future*, 4th edition. Benjamin/Cummings Publishing Company, Inc., Redwood City.

Marshall, N.B. 1979. *Developments in Deep-Sea Biology*. Blandford Press, Poole, Dorset.

Thomas, D.S.G. 1995. 'Desertification, causes and process', in W.A. Nierenberg (ed.) *Encyclopaedia of Environmental Biology*, vol. 1, 463–473.

UNEP. 1992. *World Atlas of Desertification*. Arnold, London.

Wilson, E.O. 1992. The *Diversity of Life*. Belknap Press of Harvard University, Cambridge, MA.

3 Basic ecological concepts and processes

Key concepts

- **The laws of thermodynamics determine the transfer of energy through an ecosystem.**
- **Energy flow through an ecosystem is a one-way process while other materials recycle on various timescales.**
- **Energy transfer is subject to a variety of factors and there are various measures of efficiency.**
- **The concept of trophic (feeding) levels, food chains and food webs is important for the understanding of energy transfer and loss in any ecosystem.**
- **Feeding relationships are fundamental to the organisation and function of biological communities.**
- **Every system has three main components: the autotrophic primary producers, the secondary producers (mainly the herbivores and carnivores) and the decomposers.**
- **The biosphere interacts with three main physical compartments: the hydrosphere (water), the lithosphere (land) and the atmosphere.**
- **Materials have biogeochemical cycles, the main ones being the water cycle, carbon cycle, oxygen cycle and the nutrient (particularly nitrogen, phosphorus and sulphur) cycles.**

Despite the diversity of living forms and their habitats (outlined in Chapter 2), there are a number of fundamental concepts and processes that apply to all ecosystems and the interactions of their components. Any ecosystem is a collection of interdependent components (subsystems), each with its own defined boundary. Each ecosystem has two basic requirements for its successful functioning, namely

- the acquisition, transformation and transfer of energy and
- the gathering and the recycling of the materials necessary for life.

By understanding how these systems function, environmental science is better able to understand and predict the impacts of human activities on such systems. This chapter presents an overview of the key processes involved.

3.1 Energy in ecosystems

Energy comes in many forms, such as heat, light, sound and electricity, and all have in common the capacity to do work. All forms of energy fall into one of two categories:

- *potential* (stored) energy such as that found in the high-energy chemical bonds of ATP (see Section 1.1.2.3), the main energy carrier for living things, or
- *kinetic* energy, possessed by objects in motion.

All forms of energy follow basic laws known as the Laws of Thermodynamics. Understanding these laws and their consequences will help you to understand ecology and its application to issues such as sustainability and conservation.

3.1.1 The laws of thermodynamics

The application of the two fundamental laws of thermodynamics to energy and matter transformations at the cellular level was discussed briefly in Section 1.1.2.3 but these laws apply equally at all levels of organisation.

The First Law (the law of conservation of energy) asserts that in a closed system, energy can neither be created nor destroyed but can only be transformed from one form to another. Thus, when fuel is burnt to drive a car, the potential energy contained in the chemical bonds of that fuel is converted into mechanical energy to propel the car, electrical energy to ignite the fuel, light to show where you are going and heat to defrost the windscreen. The key point, however, is that if you could measure the total amount of energy consumed and compare it with the total amounts being produced in these various other forms, the two would be equal. Energy conversions such as these also take place in biological systems. Photosynthetic organisms such as plants capture and transform light energy from the sun and transfer this energy throughout the system subject only to the consequences of the Second Law.

The Second Law asserts that disorder (entropy) in the universe is constantly increasing and that during energy conversions, energy inevitably changes to less organised and useful forms, i.e. it is degraded. Think of this as energy always going from concentrated to less concentrated forms, the least useful (i.e. least concentrated) being heat energy. The consequences of this are very significant biologically. During each conversion stage, some energy is lost as heat. Therefore, the more conversions taking place between the capture of light energy by plants and the trophic (feeding) level being considered, the less the energy available to that level. The efficiency of the transfer along food chains is generally less than 10 per cent because about 90 per cent of the available energy is lost or used at each stage (see Section 3.2.2).

3.1.2 The flow of energy

The study of energy flow is important in determining limits to food supply and the production of all biological resources. The capture of light energy and its conversion into stored chemical energy by *autotrophic* organisms provides ecosystems with their primary energy source. Most of this is *photosynthetic*, chlorophyll-based production, the exception being the comparatively limited production of organic materials by *chemosynthetic* organisms. The total amount of energy converted into organic matter is the *gross primary production* (GPP) and varies markedly between systems.

However, plants use between 15 and 70 per cent of GPP for their own maintenance. What remains is the *net primary production*, NPP (Table 2.1). The total net primary production of an ecosystem provides the energy base exploited by non-photosynthetic (**heterotrophic**) organisms as *secondary production*. Heterotrophs obtain the energy they require by consuming and digesting plants (*herbivores*), by feeding on other heterotrophs (*carnivores*) or by feeding on detritus, the dead bodies or waste materials of other organisms (*detritivores*, *saprophytes*, *saprozoites*). The energy stored in the food materials is made available through cell respiration. Chemical energy is released by burning the organic compound with oxygen using enzyme-mediated reactions within cells. This produces carbon dioxide and water as waste products.

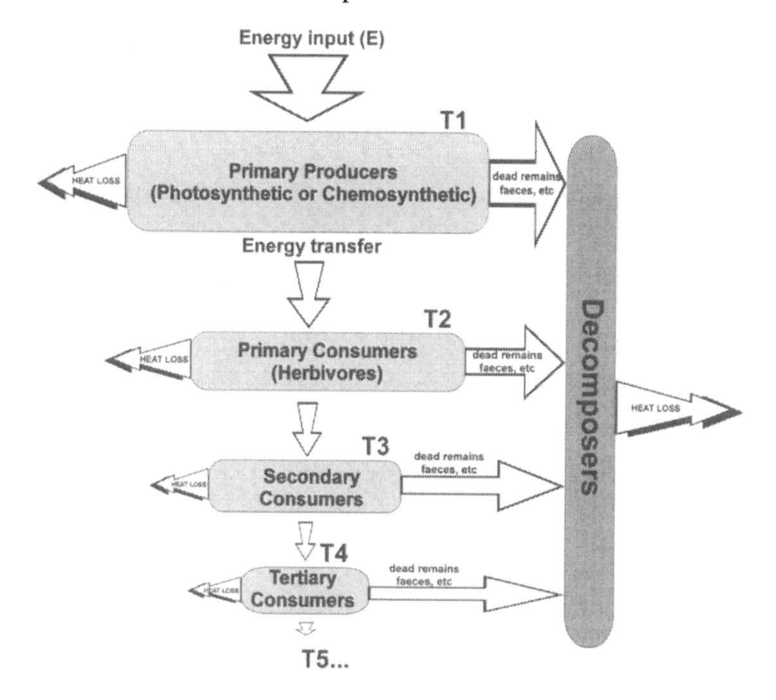

Figure 3.1 ***Energy transfer and flow through a simple hypothetical ecosystem. The area of each box is proportional to the biomass at that level***

Energy flow is the movement of energy through a system from an external source through a series of organisms and back to the environment (Figure 3.1). At each stage (trophic level) within the system, only a small fraction of the available energy is used for the production of new tissue (growth and reproduction); most is used for respiration and body maintenance. Once the importance of energy flow is appreciated, the significance of energy efficiency and transfer efficiency is more readily understood. Energy efficiency is 'the amount of useful work obtained from a unit amount of available energy' and is an important factor for the management and conservation of any biological resource. The development of most modern intensive agriculture is founded on the principle that increased channelling of energy into a system results in higher yields; however, the energy efficiency is usually less than in more traditional agricultural systems.

A common ecological measure of efficiency is the trophic-level efficiency, the ratio of production at one trophic level to that of the next lower trophic level. This is never very high and rarely exceeds 10 per cent (the '10 per cent rule'), more typical values being only 1–3 per cent. Table 3.1 lists other measures of efficiency often used in ecological comparisons. However, estimates of ecological efficiencies can vary widely between individuals and populations because individuals in a population may live under different ecological conditions.

Table 3.1 ***Ecological measures of efficiency***

Term used	*Definition*
Photosynthetic efficiency	The percentage of available sunlight that is used in photosynthesis to convert carbon dioxide into carbohydrate
Exploitation efficiency	The percentage of production at one trophic level that is ingested by the trophic level above it
Assimilation efficiency	The percentage of energy ingested that is actually absorbed by the body rather than egested
Growth efficiency	The percentage of assimilated energy that is used for growth rather than being respired or used for reproduction
Reproductive efficiency	The percentage of assimilated energy that is used for reproduction rather than being used for respiration or growth
Production efficiency	The percentage of assimilated energy that is used for production (growth and reproduction) rather than respiration
Trophic efficiency	The efficiency of energy transfer from a trophic level to the next one above

3.2 Food chains and food webs

Feeding is one of the main forms of interaction between organisms in any system. When one organism consumes another, both energy and materials are transferred. The processes involved form an important conceptual framework for describing, analysing and comparing ecosystems.

3.2.1 Feeding types and relationships

Feeding relationships are fundamental to the organisation and function of biological communities and the diet of an organism is central to the understanding of its ecology. Table 3.2 gives some of the key terms relating to trophic levels but there are many terms and classifications associated with feeding. Except for the autotrophs, which produce their food from sunlight, carbon dioxide and water, and detritivores, which consume dead material, all other organisms obtain their energy and materials from the digestion (enzymatic breakdown) of other organisms, a process termed *secondary production*. Several important groups of organisms exhibit an intimate relationship between an animal and a photosynthetic organism. This relationship is a *symbiosis*, in which both partners receive a benefit, at least one nutritionally, and it may occur in various forms. Corals (animal–alga), lichens (alga–fungus) and leguminous plants (plant–bacteria) are good examples (Box 3.1).

Some organisms are specialist feeders, feeding only on a limited range of specific organisms, while others are generalists, feeding on a diversity of organisms depending upon their relative availability. *Omnivores* are organisms that eat both plant and

Table 3.2 ***The main trophic levels***

Trophic level	*Examples*	*Energy source*
Primary producers (T1)	green plants, photosynthetic bacteria and protists; chemosynthetic bacteria	solar energy (photosynthesisers); chemical energy (chemosynthesisers)
Herbivores (T2)	cattle, elephants, rabbits, herring, locusts, most copepods, water fleas	tissues of primary producers
Primary carnivores (T3)	many fish, many insectivorous birds, foxes, lions, spiders,	herbivores
Secondary carnivores (T4)	hawks, seals, barracuda, sharks	primary carnivores
Omnivores	crabs, many birds, humans	organisms from any other trophic level
Detritivores	many bacteria, fungi, worms, millipedes, vultures, fly larvae (maggots)	dead bodies and waste products of other organisms

animal materials, humans being a good example. Some organisms also change their diet during their life cycle. The main components of an ecosystem can be classified into three categories:

3.2.1.1 Autotrophic primary producers

fall into one of two categories;

- **Photoautotrophs** use sunlight as their energy source and include the photosynthetic plants and microbes.
- **Chemoautotrophs** utilise energy derived solely from chemical reactions and include only prokaryote forms. The most ecologically important of these occur in the nitrogen cycle (see Section 3.3.4.1) and in deep-sea hydrothermal vent communities (Box 3.2).

3.2.1.2 Heterotrophic organisms,

the secondary producers, can usefully be grouped into four main functional types:

- True *predators* kill their prey soon after attacking them and consume several or many different prey individuals over their lifetime; they may or may not consume the prey in its entirety. They include the obvious carnivores such as lions, sharks and birds of prey, but also carnivorous plants, whales of all types and many filter-feeding forms. Seed-eating birds and rodents could also be classed as true predators since they consume future individuals (Box 3.3).

Box 3.1

Symbiosis: fair exchange is no robbery

Symbiosis (meaning 'living together') is the intimate association between members of two species in which no harm is caused to the other partner. In *mutualism*, both partners benefit from the association, whilst in *commensalism*, only one partner benefits. In commensal relationships, the advantage gained by one species often involves the provision of shelter, support, transport and/or food. Good examples are epiphytic plants and clown fish that gain both protection and some food by living amongst the tentacles of large sea anemones in the tropics. However, the more spectacular relationships are those of mutualistic associations, which are quite widespread.

Mutualistic relationships exist between

- plants and micro-organisms, e.g. leguminous plants such as peas and beans protect and encourage the growth of nitrogen-fixing bacteria (*Rhizobium* spp.) in their root nodules and receive usable nitrogen compounds in return. Such plant systems can be very important in nitrogen-poor soils.
- protists and fungi, which come together to form compound organisms called lichens. They contain highly modified fungi in association with either cyanobacteria or green algae. The fungi provide a supporting structure and absorb water and nutrients while the micro-organisms photosynthesise and provide the chemical products for the fungus.
- terrestrial plants and insects, intimately associated through pollination mechanisms, e.g. bee orchids. *Acacia cornigera* trees provide special facilities for a particular genus of ants (*Pseudomyrmex*) which in return protect the tree from many leaf-eating insects and other herbivorous species.
- animals and protists, which have extremely important associations. Herbivores such as termites and ruminants use gut bacteria to enable them to digest cellulose. Corals and giant clams culture algae within and between their cells, providing the algae with protection and nutrients in return for photosynthetic products.
- animals and other animals. For example, some ants culture and raise aphids in return for sweet secretions produced during aphid feeding. Cleaner fish and shrimps remove ectoparasites (as food) from their customers, whose inaccessible regions are protected, therefore, from these parasites.

Some of the most important mutualisms occur between different kingdoms and are of great ecological significance, e.g. coral reefs are founded upon a mutualism between the coral and unicellular algae.

- *Grazers* attack large numbers of prey (plant or animal) also but they may remove only a part (e.g. shoot or limb) of each prey individual. Their impact is usually damaging but not lethal in the short term. The most obvious examples are the large grazing herbivores such as elephants, deer, sheep and cattle but this category could also include organisms such as biting flies and leeches.
- *Parasites*, organisms living in or on their prey and obtaining their nutritional needs from another living organism, consuming only parts of their prey. Their attacks are usually harmful but rarely lethal in the short term since this would not usually be in the parasite's interests. Unlike grazers, their attacks focus on only one or a few individuals during their lifetime, resulting in the development of an intimate

Box 3.2

Hydrothermal vent communities: chemosynthetic oases on the deep-sea floor

These remarkable systems were discovered in the late 1970s by scientists diving at 2700 m on the Galapagos rift zone in the submersible *Alvin*. They revealed a spectacular concentration of previously unknown animals on the otherwise barren deep ocean floor. The most prominent animals were unusually large tube worms (*Riftia*) more than a metre long, clams up to 25 cm long, large mussels and two species of white, blind crabs. The hydrothermal activity results in the release of mineral-rich, hot water into the sea water. Diffuse flows from cracks and crevices in the sea floor are at temperatures up to 100°C, while chimney-like vents (black smokers) release superheated water at temperatures as high as 400°C. The result is a general temperature of 8–16°C in an area normally at only about 2°C. The water also contains high concentrations of hydrogen sulphide (H_2S), which can be used as an energy source by chemosynthetic, sulphur-oxidising bacteria (e.g. *Beggiatoa* and *Thiomicrospira*). These organisms use the same basic biochemical pathway as photosynthetic organisms. Chemosynthetic production in the vents is summarised as:

$$CO_2 + H_2S + O_2 + H_2O \rightarrow H_2SO_4 + CH_2O \text{ (i.e. a carbohydrate)}$$

These primary-producing bacteria form the basis of the food web in this unique community, which is totally independent of sunlight, the system being driven entirely by geothermal energy. Bacterial production here is estimated to be two or three times greater than that of the photosynthetic primary producers in the pelagic waters above. These bacteria either

- are eaten directly by heterotrophs: the bacteria form dense mats up to 3 cm thick that can be grazed;
- live symbiotically with the larger animals (e.g. the clam *Calyptogena* sp. and the vestimentiferan worms *Riftia* spp.);
- die to form particulate organic matter (detritus) used by detritivores and filter feeders.

Many of the animals in these densely populated communities are surprisingly large with an atypically large biomass (*Riftia* alone up to 9100 g wet weight/m^{-2}: community up to about 30 kg wet weight m^{-2}) for deep-sea benthos.

Since the discovery of the hydrothermal vent communities, further examples of chemosynthetically based communities on the sea floor have been discovered, all based on the release of sulphide-rich water. Temperature does not seem to be a critical factor since cold-water seeps have been found with communities essentially similar to those of the hydrothermal vents. These communities are called shallow vents and cold seeps.

relationship between the host and the parasite. They include tapeworms, liver flukes, aphids, plants such as mistletoe and considerable numbers of fungi (e.g. rusts) and micro-organisms (viral and bacterial diseases).

- *Parasitoids* are a specialist group of insects belonging to the Hymenoptera and Diptera. They lay eggs in or near other insect hosts which the parasitoids devour during larval development. They may seem to be a group of limited general importance but it has been estimated that they account for 25 per cent of the world's insect species. Their specialisations are the source of much ecological investigation.

Box 3.3

Malacosteus: a deep-sea predator extraordinaire!

The *mesopelagic* zone is an area of remarkable specialisations such as the widespread use of bioluminescence (see Box 2.2). The privation encountered in areas of low faunal density makes carnivory potentially more difficult since organisms meet prey relatively infrequently. This has led to a variety of adaptations such as large teeth and elastic stomachs to allow the capture of comparatively large prey items. One of the most spectacular animals of this zone is the stomiatoid fish, *Malacosteus* sp. (Figure B3.1) also known as the rat-trap fish, which grows to about 15 cm in length and is a highly adapted and voracious predator.

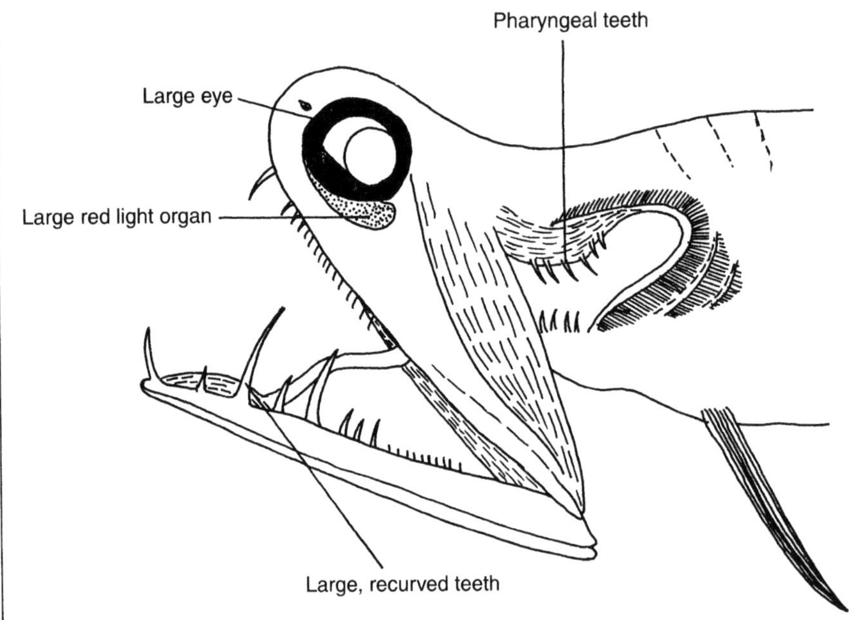

Figure B3.1 ***The head of*** **Malacosteus** ***sp. (the rat-trap fish), a highly adapted deep-sea predator***

The following are key adaptations shown by this species:

- The development of large, inwardly curved teeth whose action is such that any movement of the prey can only result in aiding ingestion of the prey.
- The loss of the skin around the jaw mechanism to facilitate rapid 'lunging' movements of the fish during the act of capturing prey. These fish appear to hunt by creeping up on prey with the mouth held open like a trap. Once within reach, they snap the jaws rapidly forward. The loss of skin from around the jaw mechanism allows a more rapid movement through the viscous watery medium.
- The reduction of the muscles required to open the jaws also streamlines the jaw mechanism. The closing muscles remain relatively large and powerful to provide a firm hold on the prey item.
- Because they need to ingest large prey items, they have a problem of potential suffocation. Fish typically pump water over their gills using a mouth mechanism which becomes unavailable during the slow ingestion of a large prey item. This fish, therefore, has externalised its gills so that it can respire passively while engaged in ingesting prey items.

Box 3.3 continued

- The connection between the head and body has been minimised to just a thin 'neck'. This facilitates the snapping mechanism and reduces the transmission of shocks down the body while a struggling prey item is being held on to. It also means that the pharynx does not directly connect to the jaws and prey items need directing into the pharynx by the development of special pharyngeal teeth.
- A large red light organ beneath each eye produces a beam of red light for hunting prey (mainly large, red crustaceans). This is particularly significant since it is considered unlikely that the potential prey will be able to perceive red light (because red light is not normally present at this depth). They are, effectively, being hunted by an invisible beam of light.

The key point about predation and herbivory is that when energy and materials transfer from one trophic level to another, the prey population suffers to a degree that depends upon the intensity and the type of predation. Relative size and strength of both predator and prey influence the diet of organisms that hunt and kill while for passive, filter-feeding organisms the size of the food particles is important. An important corollary of predation is the evolution of protective mechanisms to reduce predation: these include morphological, chemical and behavioural mechanisms.

3.2.1.3 The decomposers

The most important organisms in any community are the primary producers, without which there can be no life. However, the decomposers are equally vital since it is their activities that recycle the materials necessary to allow the primary producers to continue producing. Without decomposers, the remains of the primary and secondary producers would accumulate, tying up the substances in the dead bodies and preventing further production. Decomposers break down the dead remains and waste products to release the materials vital for life (Figure 3.2)

Decomposition is essentially a biological process, aided by a variety of physical processes such as weathering and leaching. The rate of decomposition varies spatially and temporally, influenced by a number of factors, including:

- the type of detritus (plant and animal debris),
- the type and abundance of decomposer organisms present, and
- environmental conditions such as temperature, moisture, aeration and nutrient availability.

However, the decomposers are largely aerobic species requiring oxygen to degrade the organic materials. Aeration is, therefore, very important for the decomposition process. By stimulating decomposition activity, large loadings of organic matter (such as sewage) can markedly deplete oxygen levels in aquatic systems, resulting in a deterioration of their ecological and amenity values.

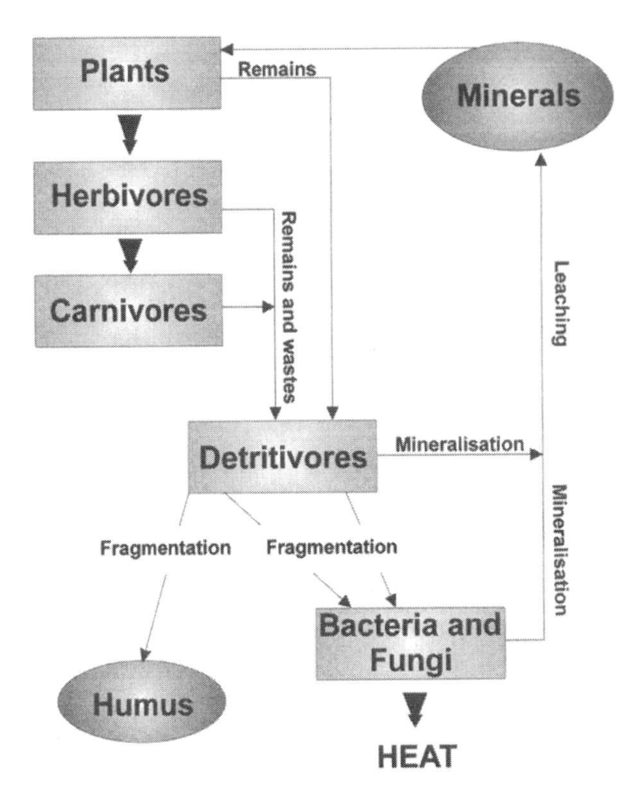

Figure 3.2 ***A summary of the role of decomposers (detritivores and bacteria/fungi) in an ecosystem***

Most decomposers are micro-organisms and decomposers have been neglected in most studies of natural communities. They are heterotrophic organisms that can also be termed *saprotrophs* because they feed on dead organic matter. They may also be called detritivores since they break down plant and animal debris (detritus). When the detritus originates from within the system being considered it is termed **autochthonous** detritus but when derived from outside the system, it is termed **allochthonous**. Allochthonous detritus is very important in aquatic systems such as estuaries, rivers and streams where primary production is restricted. Organic matter associated with soil is called humus and differences in *humus* types characterise different soil types.

Most decomposition is carried out by fungi, bacteria and by invertebrates and organisms of a variety of sizes necessary for optimum decomposition rates, each having its role in the total breakdown sequence (Box 3.4). Decomposition also includes the process of *mineralisation*, the conversion of mineral elements from an unavailable organic form to an available inorganic form.

3.2.2 Trophic levels

Energy, chemical elements and some organic compounds are transferred from organism to organism through a sequence frequently described as a *food chain* (Figure 3.3a). However, such linear structures are rare and complex *food webs* (Figure 3.3b) are more typical. Organisms within food chains/webs can be organised into *trophic levels* (Figure 3.3 a, b and Table 3.2). Each trophic level comprises all organisms which are the same number of feeding levels away from the original energy source (usually the sun). This particular concept has a number of difficulties, particularly those of organisms feeding at more than one trophic level and the problems associated with the trophic position of various decomposers. However, it is widely used and does illustrate a number of important points relating to the flow of energy and materials through the ecosystem.

At the base of the chain (the primary producer level or first trophic level T1) is always a green plant or other autotroph capturing energy and converting it into energy-rich

Box 3.4

Breakdown of leaf litter: decomposition at work

Decomposition is as vital to life as is primary production, although without the latter there could be no decomposition. Without decomposition, the biosphere would quickly grind to a halt because resources such as nutrients would become tied up in the bodies of dead organisms. Studies of natural communities often pay inadequate attention to the decomposers, perhaps because of their usually small size, most being bacteria, saprophytic fungi and invertebrates.

In most forests, leaf litter does not accumulate but decomposes at a similar rate to that of deposition although this may be subject to seasonal variations in temperate environments. The amount of material accumulated is generally greater at higher latitudes (Figures B3.2 and B3.3). In tropical rainforests, almost all the organic matter is tied up in the living organisms and organisms depend upon a rapid turnover of nutrients for their survival. Breakdown in the tropics is rapid because the hot and humid climate provides an ideal environment for diverse decomposing organisms. Experiments have shown that decomposition requires the presence of a wide range of organisms of varying size and types. Although micro-organisms are clearly vital to the process, the presence of soil macrofauna such as earthworms, woodlice and millipedes is important in facilitating their activities. The rate of decomposition of leaf litter and release of nutrients varies widely. Some forms of litter such as wood, conifer needles and *sclerophyllous* leaves are more resistant to decomposition than broadleaf deciduous and tropical rainforest leaves. Decay rate is broadly correlated with temperature, being slow in colder climates and faster in warm ones. Half-times for decomposition of leaf litter exceed ten years in northern coniferous forests; several years in southern pine forests and temperate deciduous forests but only months in tropical rainforests.

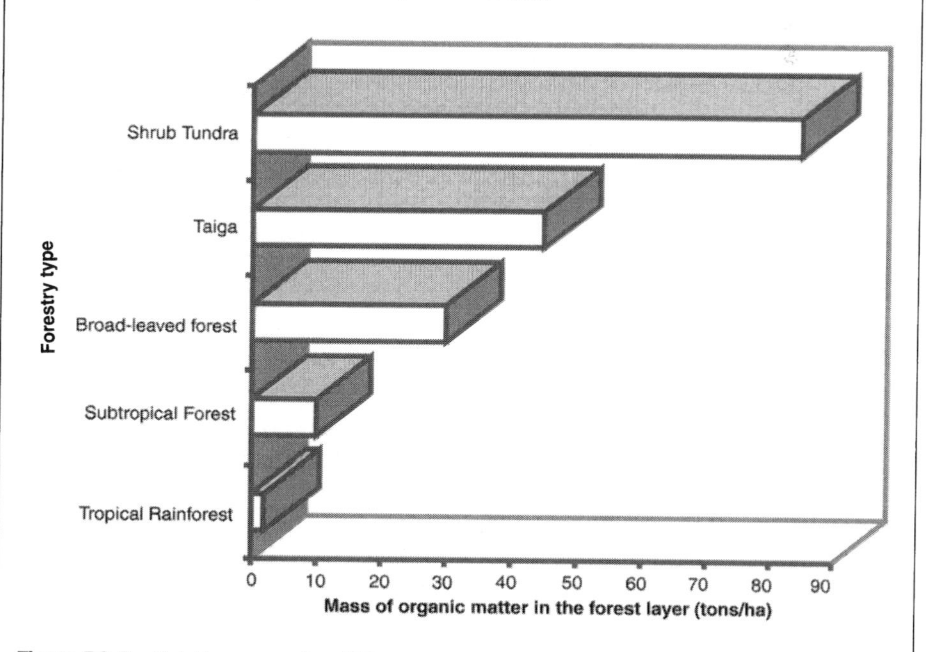

Figure B3.2 ***Relative amounts of litter accumulated in the forest layer in forests from different latitudes and habitats***

Box 3.4 continued

When organic materials decay, they undergo humification to produce humus, organic material derived from partially decomposed material in the soil and an important component of fertile soils. The colonisation of dead plant tissue is a successional sequence (see Section 7.4) where both physical and chemical degradation processes take place.

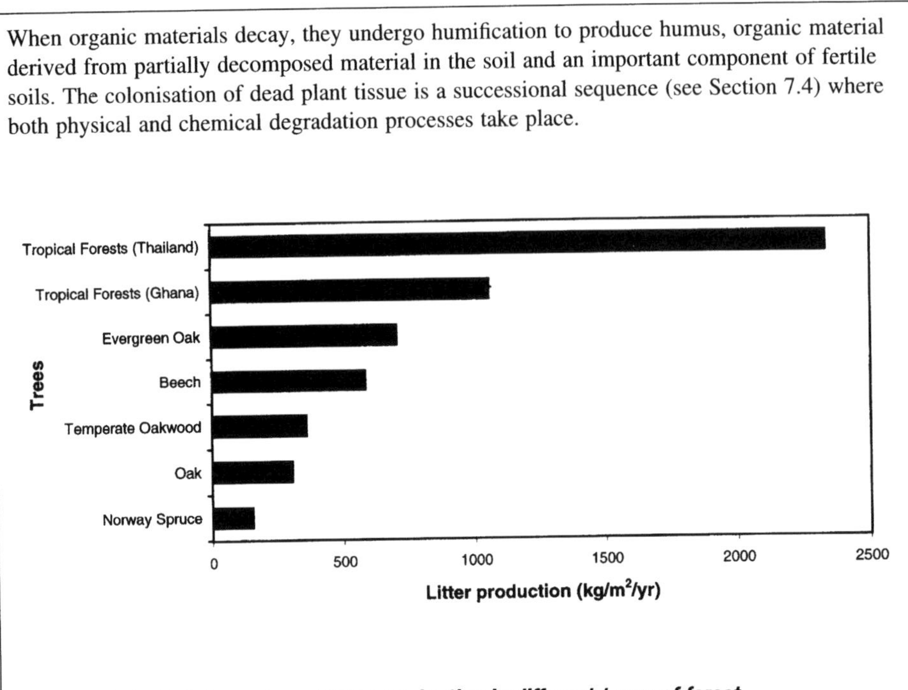

Figure B3.3 ***Relative rates of litter production in different types of forest***

compounds. These are then available to the next trophic level, the primary consumer or herbivore level (T2). The body of the herbivore then provides the energy source for the next trophic level, the secondary consumer or primary carnivore level (T3). A variable number of further consumer levels may then occur (T4–n), the final ones always being detritivores, organisms which consume dead material. Some organisms such as humans may obtain their food from more than one trophic level within a food web (omnivores).

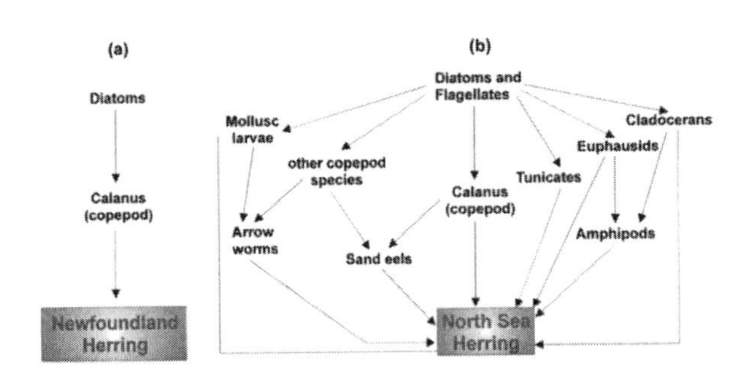

Figure 3.3 ***A comparison of a food chain (a) with a food web (b) for related species of herring. The Newfoundland herring feeds on a particular species of copepod while the North Sea herring has substitutable food resources***

The conversion efficiency from one trophic level to the next is poor, about 90 per cent of the available energy being lost during each transfer. The average value transferred is about 10 per cent although it ranges between 5 and 20 per cent in various systems. Figure 3.4 illustrates the reasons for this low conversion efficiency and includes the following features:

- First, not all material at any trophic level is eaten.
- Second, not all material eaten is digested, much passing through the digestive system unchanged and being **egested** as roughage.

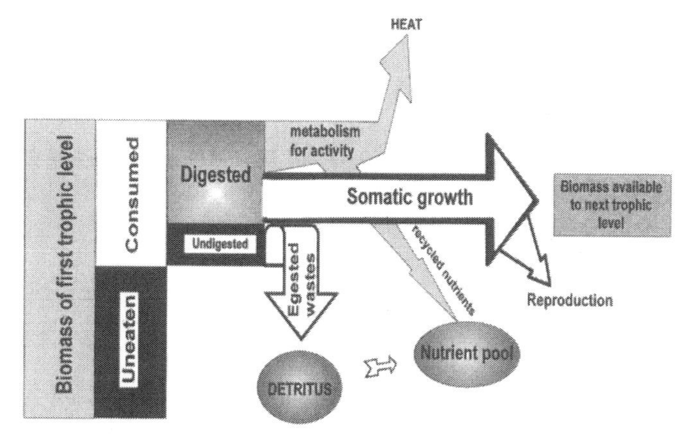

Figure 3.4 ***A summary of the ways in which energy is lost during trophic level transfers***

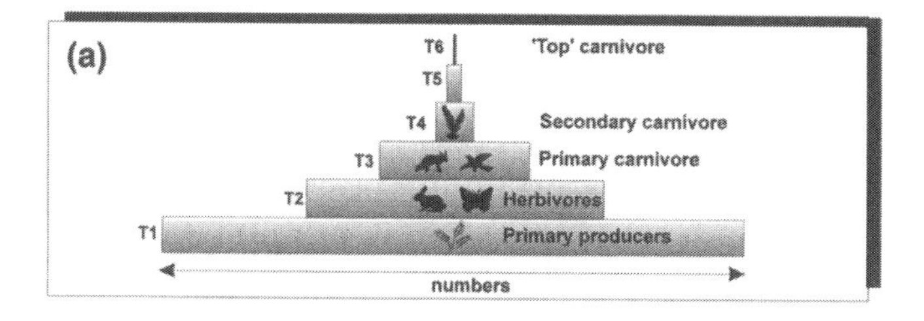

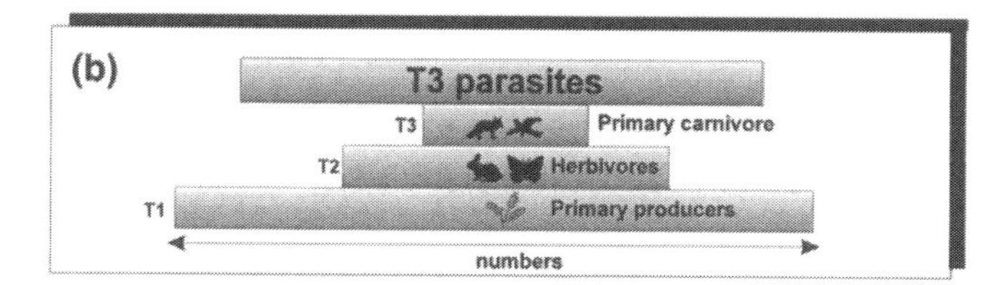

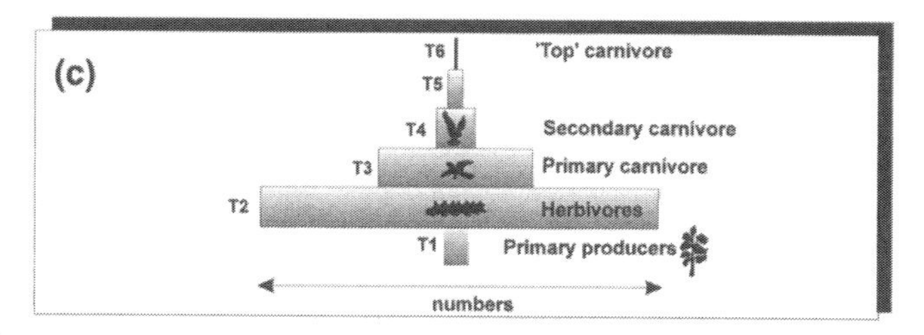

Figure 3.5 ***Examples of pyramids of numbers. (a) The typical theoretical pyramid showing a steady decline in abundance with increasing trophic level. (b) the inverted pyramid associated with a parasitic trophic level. (c) The pyramid associated with the occupation of a large food resource such as a tree where the abundance of the primary producer obscures its supporting role (in primary production terms)***

- Lastly, most of the material digested is used for catabolic processes such as respiration, which provides the energy needed to maintain the organism.

Only a small amount of energy remains available, therefore, for transfer to the next trophic level. This is particularly important since it places limits on the potential number of trophic levels, few food chains having more than four trophic levels. It also explains why there are few big, fierce animals, the energy available to higher trophic levels rarely being sufficient to maintain large viable populations of high-level consumers. Top predators such as lions or great white sharks are not themselves preyed upon as food sources.

3.2.3 Ecological pyramids and the flow of energy

The need to quantify the relationships between trophic levels led Elton (1927) to introduce the concept of the pyramid of numbers. Figure 3.5 shows the relative abundance of individuals at each trophic level. The quantification of trophic relationships is usefully presented as pyramids of biomass and energy also, each pyramid providing a different type of information about the relationships between the members of a community. The main ones are discussed below.

3.2.3.1 Pyramids of numbers

Abundance is usually measured as number of individuals per unit area. Because of the energetic factors described in Section 3.1.2, the number of individuals present usually declines at each trophic level, forming a pyramid of numbers (Figure 3.5a). In practice, it is very difficult to determine the actual number of individuals at each trophic level in a food web. The two most common exceptions to this rule are:

- the inverted pyramids associated with parasites (Figure 3.5b), where each top carnivore may carry large numbers of parasites, and
- pyramids associated with large primary producers such as trees (Figure 3.5c), each of which may support large numbers of smaller organisms.

3.2.3.2 Pyramids of biomass

The number of organisms supported at each trophic level is size-related and so biomass is the ecologists preferred simple measure of abundance at each trophic level. *Biomass* is the total dry weight of organic matter present at each trophic level and is usually expressed as a dry weight per unit area or volume. Because water content varies between organisms and is not 'produced' by the organism, water is excluded from this measure. The '10 per cent rule' described earlier means that pyramids of biomass (Figure 3.6a) are not usually inverted. However, because measures of biomass are instantaneous measures (*standing crop*) of a dynamic situation more properly measured by productivity (see Section 7.3), some pyramids can appear to be inverted. Notable examples are those describing the marine plankton (Figure 3.6b), where a given standing crop of phytoplankton appears at some times of year to be supporting a larger standing crop of zooplankton. However, this is a consequence of the phyto-plankton reproducing at such a rate that their productivity actually exceeds that of the zoo-plankton. It is, therefore, the pyramid of productivity that

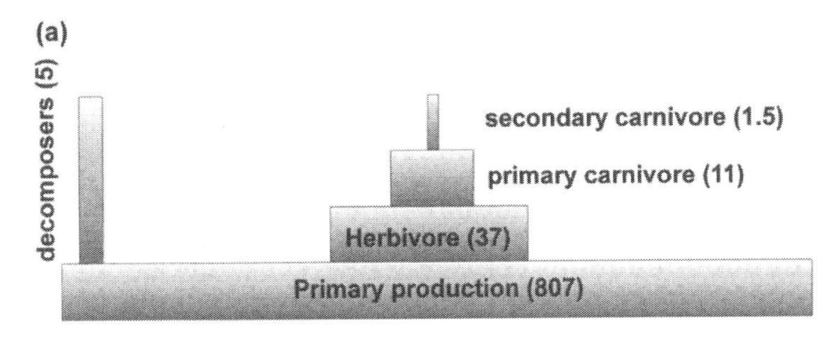

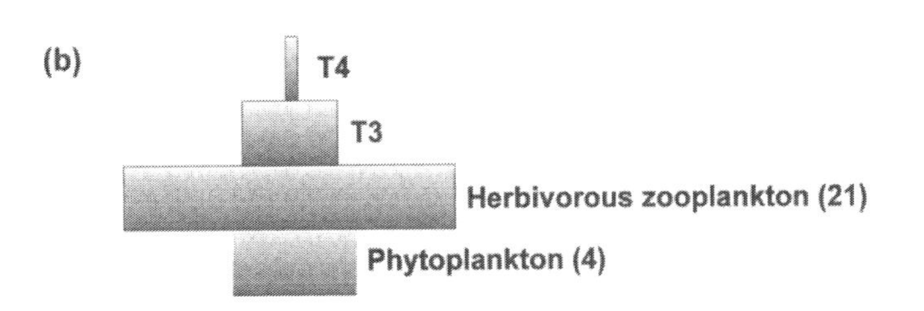

Figure 3.6 ***Pyramids of biomass in which all quoted values are in g per sq. m. (a) The expected pyramid resulting from the 10 per cent rule. (b) The inverted pyramid typical of marine planktonic food webs. This occurs because the extremely high production rates of the phytoplankton enable a relatively small standing crop to support a much larger biomass of zooplankton***

is of true ecological significance, pyramids of numbers and biomass being less reliable, but more simply measured, derivatives of it.

3.2.3.3 Pyramids of energy flow/productivity

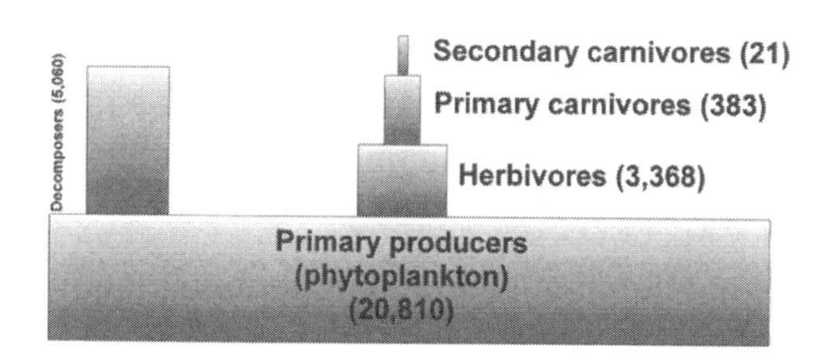

Figure 3.7 ***Pyramid of energy flow for an aquatic system in Silver Springs, Florida. Values are kcal per sq. m per year (adapted from Miller 1994)***

These show the energy present at the different trophic levels of a community, measured as energy per unit area per unit time, e.g. kJ/ha/yr. They illustrate, therefore, the rates of energy flow through food webs (Figure 3.7). They can never be inverted because of the First Law of Thermodynamics (the law of conservation of energy) and the productivity of successive trophic levels must be less than that of the level below.

3.2.4 Food chains and food webs

A *food chain* (Figure 3.8) illustrates the transfer of matter and energy from primary producer to decomposer in a particular habitat. It is rare for a food chain to have more than five trophic levels because of the low efficiency of energy transfer from level to level (Section 3.2.2) and because generally, the longer the chain, the less stable it will be: there are more links to fail. The shorter the chain, the more effectively primary production is utilised; it is therefore more efficient energetically to be a herbivore. However, true food chains are rare in nature and have inherent risks due to the specialisation involved (Box 3.5). Few data on the length of food chains exist but, since they are anyway largely hypothetical, they provide only limited information about the way communities really operate. Most organisms feed, or are fed upon, by more than one organism and thus the linear structure of a chain is artificial. A more realistic representation of feeding relationships for most systems, therefore, is the *food web*.

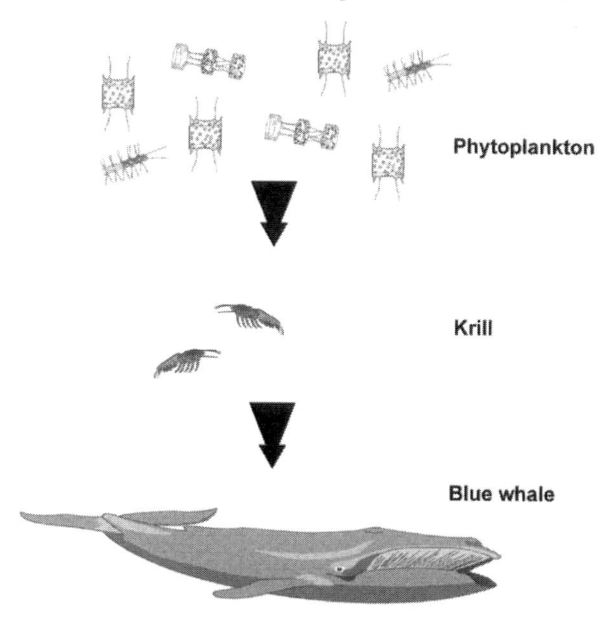

Figure 3.8 ***The food chain of the blue whale, an example of a short, efficient trophic system based upon extreme specialisation upon a single food resource***

Box 3.5

The dangers of feeding specialisation: chains = risky, webs = safety

Linear food chains are relatively uncommon in nature, while most organisms are part of a food web. Why? At least part of the answer lies in the very long-term risks associated with narrow specialisation. Linear food chains are seen in the Newfoundland herring (Figure 3.3a) and the blue whale (*Mysticetus mysticetus*) (Figure 3.8). Both of these chains are short (only three trophic levels) and, therefore, energetically efficient but the herbivore level comprises only one particular organism upon which the next trophic level is totally dependent. This is not a problem when the relationship has evolved because of long-term stability and abundance of the herbivore populations and those populations remain stable. However, potentially catastrophic declines are probable if disease or some other factor results in the decline of the lower trophic levels.

The clearest example of this is shown by the blue whale, which is totally dependent upon the krill (*Euphausia superba*) populations which abound in the waters around Antarctica and have done so for geological periods of time. Humans, having exploited the blue whales by hunting them virtually to the point of extinction, eventually protected them to allow the long process of regeneration of the populations to take place. However, in the intervening period, some nations have identified the massive populations of krill as a potentially valuable source of human food. They therefore began removing large quantities of krill from the oceans and are now threatening the whale's recovery by diminishing its food supply. Because of its specialisation, the whale cannot use a substitute food source, and this places it in direct competition with the human population.

Being part of a food web means that there are substitutable food resources so that, in a year when species X is in short supply, species Y can be eaten instead. This is clearly a safety mechanism used by many species to ensure the survival of the species in the face of the dramatic natural fluctuations common in many prey populations forming the lower levels of food webs.

A food web (Figure 3.9) is effectively a matrix of food chains showing the patterns of energy and material flow through a community. It is more realistic because, for example, it can include omnivores. However, it too has its limitations. Food web diagrams rarely show the relative importance of the different links. Some links may represent 80 per cent of a diet, others only 20 per cent, but they are rarely distinguished. Their complexity is also a disadvantage and as a result they often contain varying degrees of detail, some components being identified to species level while others are just generalised groups (e.g. pelagic fish). A more meaningful summary, but one which is much more difficult to construct, is an energy-flow diagram (Figure 3.10) which does indicate the relative importance of the different relationships but requires much more effort to produce.

Important features associated with the functioning of food chains/webs are the processes of *bioconcentration* and *biomagnification* of materials such as pollutants along the trophic system. At any trophic level, organisms have the potential to concentrate materials from the environment in significant amounts, directly or indirectly, through feeding or respiratory activities. This is bioconcentration and leads

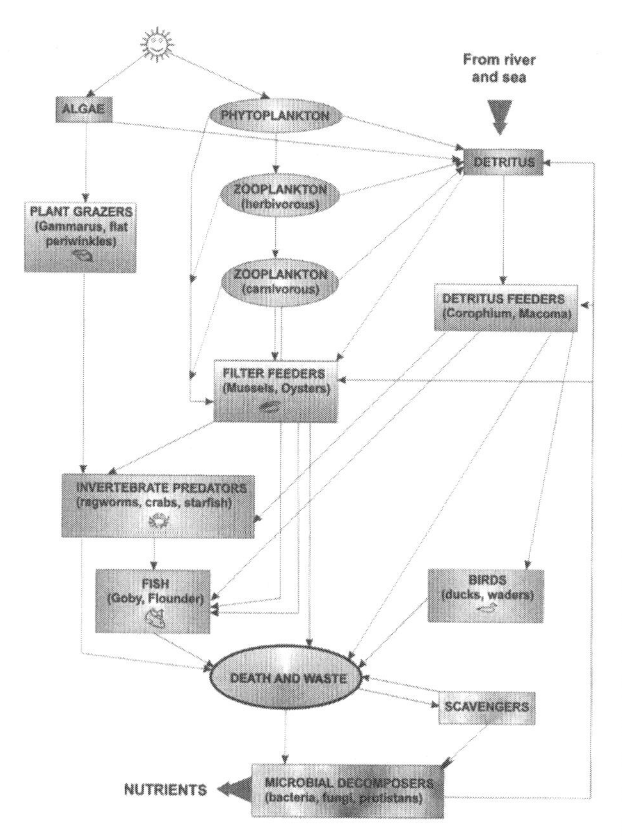

Figure 3.9 ***A typical estuarine food web showing the dominance of substitutable food resources at many levels***

to the concentration of toxic substances from the environment to dangerous levels, e.g. toxins from dinoflagellate blooms in the oceans cause paralytic shellfish poisoning in humans (Box 3.6).

When one contaminated organism consumes another, some materials can be further concentrated during their passage through the food web, a process termed biomagnification: such materials become increasingly concentrated at each trophic level. This process has been well documented for heavy metals such as mercury in both freshwater and marine systems and for the pesticide DDT (Box 3.7). The combination of bioconcentration and biomagnification along food chains/webs poses very significant threats to many ecosystems and to their exploitation as human food resources (e.g. Carsen 1962; Colborn, Myers and Dumanoski 1996).

3.3 Biogeochemical cycles and material recycling

Energy flow through the ecosystem in essentially a one-way process since the energy is increasingly dissipated as heat at each trophic transformation. However, the materials required for life are not lost as they are incorporated into new trophic levels since these materials are recycled subject to a variety of environmental constraints and timescales. Thus a variety of *biogeochemical cycles* occur for the materials required by living organisms. Note, however, that not only nutrients are subject to such cycling; pollutants such as heavy metals also have such cycles. Every element has a somewhat different fate depending upon its physico-chemical properties, its natural distributions and its biological role. The recycling process involves the movement of atoms and molecules between organisms and their physico-chemical environment. The study of such movements requires consideration of:

- the chemical forms (chemical species) involved in storage and transference,
- the processes by which they are transferred and
- the nature of any chemical transformations that occur during this process.

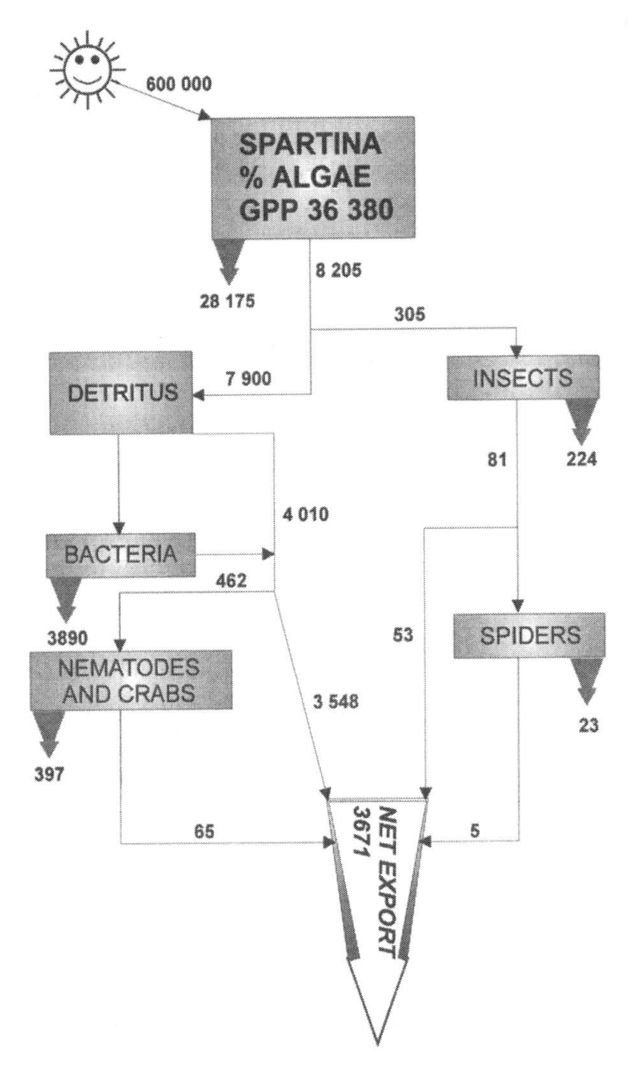

Figure 3.10 ***A simplified energy-flow diagram for a Georgia saltmarsh. The broad arrows indicate losses from the system. Values are kcal per sq. m per year (adapted from Teal 1962)***

Biogeochemical circulation occurs on a variety of spatial and temporal scales and six major biogeochemical cycles are particularly important to life, namely those of carbon, nitrogen, oxygen, phosphorus, sulphur and water. All but phosphorus have gaseous components and so cycle over large distances with relative ease. Phosphorus is completely non-gaseous and, therefore, cycles only within much more limited spatial scales. The timescale for circulation depends upon the fate of individual atoms/molecules. If they become tied up in sediments and rocks, they may become unavailable to organisms for millions of years. Molecules entrained within the food web cycle may circulate relatively rapidly until they are lost from the short-term rapid cycles to the long-term geological timescale components of the biogeochemical cycle.

To understand the biogeochemical cycling of materials, it is useful to consider the global system as comprising four distinct but interacting compartments (Figure 3.11), namely:

- the oceans and the fresh waters (**hydrosphere**),
- the land (**lithosphere**),
- the **atmosphere** and
- the **biosphere**.

The physical environments and the types of organisms living there are different, as are the processes which affect the rates of circulation to, from and within these compartments.

- **Oceans** are deep (average depth of about 4000 m) and contain enormous volumes of water and associated materials. Exchange of materials, however, takes place mainly at the surface and at their edges, where there is runoff from the land through rivers and estuaries. Because of this, oceans respond very slowly to changes on land and in the atmosphere and only in the shallow water over continental shelves (neritic water) does mixing occur relatively quickly. Oceans are the ultimate store for most materials on very long timescales, deep-ocean sea-water being hundreds of years old. Most elements entering the oceans eventually settle to

Box 3.6

Toxic algal blooms: an increasing threat?

Toxic algal blooms occur in many water bodies and, although they are not a new phenomenon, the scale and complexity of these events appears to be growing. The number of toxic blooms, the economic losses resulting from them and the kinds of toxins and toxic species have all increased

In freshwater systems, the main group causing toxicity is the planktonic blue-green algae (Cyanobacteria). Some of these can produce alkaloid neurotoxins which block neuromuscular junctions; peptide hepatotoxins which cause liver damage; and/or lipopolysaccharides that mainly cause skin irritations. The purified hepatotoxins rank somewhere between cobra venom and curare in toxicity, with lethal doses being between 30 and 300 μg per kg bodyweight, about one hundred times more toxic than cyanide. The main nuisance species are members of the genera *Anabaena*, *Microcystis* and *Aphanizomenon*.

In marine systems, it is single-celled planktonic algae that are the main toxic agents when they occur as vast blooms. They often occur in sufficient quantities to colour the water, creating 'red tides' (which may actually be red, green, brown or even colourless), a name given to diverse lethal phenomena of this type. Only a few dozen of the diverse marine phyto-planktonic species are known to be toxic, mainly being dinoflagellates, prymnesiophytes or chloromonads. Several different modes of killing may result from the production of intense blooms. Paralytic shellfish poisoning results from human ingestion of shellfish which have accumulated algal toxins through their normal feeding activities: the shellfish themselves are rarely affected. The dinoflagellate *Gymnodinium breve* produces a neurotoxin that frequently causes major fish kills; when fish swim through the bloom the algae rupture and release the toxin, which causes asphyxiation of the fish.

The dynamics of such algal blooms are poorly understood. Cyanobacteria may be toxic at some times and not at others. The reasons for the formation of algal blooms are poorly understood also but have both physical (currents, temperature and stratification) and biological (grazing pressure and reproductive rates) components. One of the common factors does appear to be nutrient enrichment of a water body and both red tides and cyanobacterial blooms appear to be at least partially correlated with increases in nutrient concentrations.

It is generally considered that the Cyanobacteria are particularly likely to pose increasing public health problems for freshwater supplies.

the bottom and remain there until raised to the surface, usually as sedimentary rock, by geological events taking millions of years.

- **Fresh waters** are much smaller bodies of water than marine ones and elements are not usually locked up in the sediments for such long periods of time. They receive material from rainfall but most comes from the weathering of rocks, material being carried into the system as groundwater or surface runoff. Lakes exhibit annual cycles that depend upon depth, volume and climatic patterns, with stratification (Figure 2.4, Section 2.2.2.2) of the water body an important factor determining the rate of flux of the nutrients. Deep tropical lakes stratify strongly, so that the deeper layers become permanently deoxygenated through decomposition and restricted water circulation.

Box 3.7

The warnings of the DDT story

DDT (**D**ichloro**d**iphenyl**t**richloroethane) is the best known of a number of chlorine-containing pesticides. It was used extensively in the 1940s and 1950s to kill malarial mosquitoes and the lice that spread typhus. As such, it saved millions of lives. Unfortunately, experience proved that DDT had a number of unfortunate and unpredicted ecological consequences, so much so that its use was banned in the USA in 1972 and by 1980 its export from the USA ceased. However, its manufacture and use in less developed countries continues. The main lessons learnt from experience with DDT are:

- DDT is *persistent*, being broken down only very slowly by decomposers. It has a *half-life*, the time taken for half of a substance to degrade, of about three years. This results in environmental accumulation and distribution of the substance. DDT is now found everywhere in the world from the Antarctic ice to the bodies of humans.

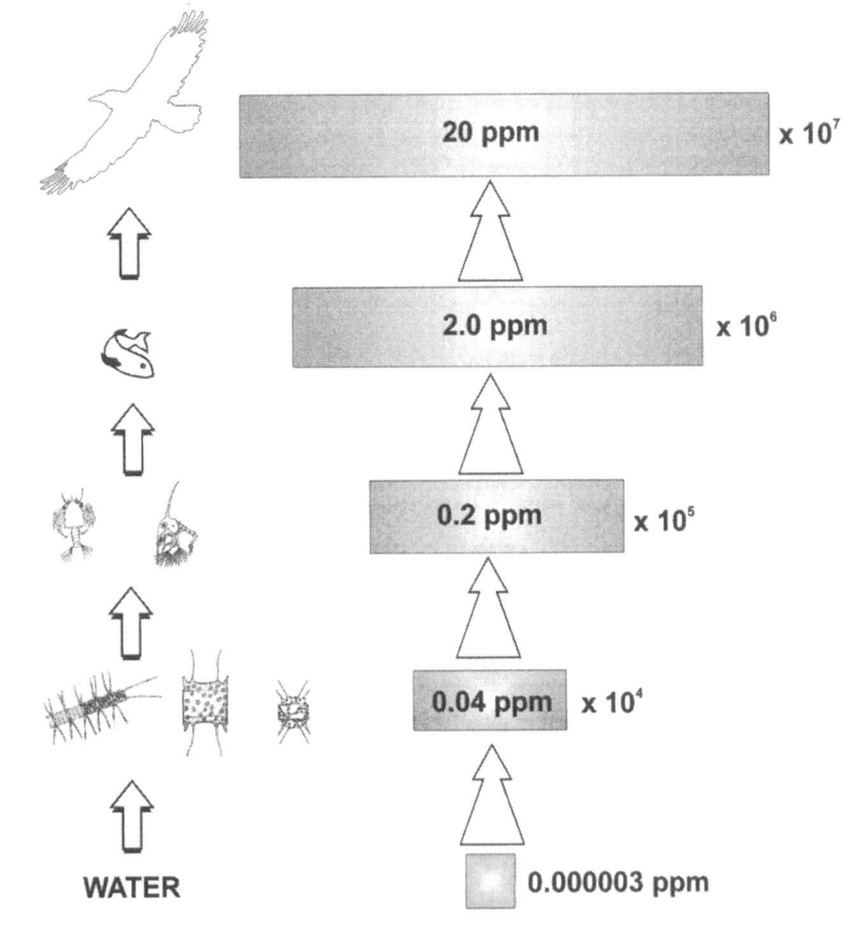

Figure B3.4 ***Biomagnification of DDT along an aquatic food chain. The concentration factors (relative to that in the water) are shown alongside the boxes. The levels in the top predator may be 7 million times greater***

Box 3.7 continued

- DDT was the first pesticide to which diverse insect pests developed a *resistance* by evolutionary selection for resistant genotypes in exposed populations. Thirty-four species of the malaria-carrying *Anopheles* mosquito are known to be DDT-resistant.
- The primary breakdown product is DDE (**D**ichlorodiphenyl**d**ichloro**e**thane), produced by dechlorination reactions that occur in alkaline environments or enzymatically in organisms. Unfortunately, DDE is almost as persistent as DDT and is responsible for shell-thinning in predatory birds. Breakdown products must always be considered when the risks associated with chemical releases are being evaluated.
- Bioaccumulation, the retention or building up of non-biodegradable or slowly biodegradable chemicals in the body to produce what is termed a body-burden of a substance, is an important process, especially for persistent materials. DDT is particularly soluble in lipids but not very soluble in water (less than 0.1 ppm) and both it and its metabolite DDE readily accumulate in the fatty body reserves of organisms.

The problem of bioaccumulation is compounded in aquatically based food chains by *biomagnification* (biological amplification), whereby the concentration of a chemical increases at each trophic transfer. This process can result in concentration factors between trophic levels of one or two orders of magnitude and for heavy metals such as mercury, factors of a thousand times have been reported. (Figure B3.4). Concentration factors along the entire food chain may result in the top predator containing several million times the concentrations in the water column.

Although it was clear that high concentrations of DDT would be lethal, other *sub-lethal* mechanisms of ecological and public health impact have become evident. These include a reduction in resistance to diseases, parasites and predators, and a reduction in reproductive capability. Populations of top predator birds such as fish-eating ospreys and bald eagles and rabbit-eating birds of prey declined dramatically in the 1960s. This was not only a consequence of deaths resulting from lethal concentrations of DDT/DDE being accumulated but also because of loss of eggs due to eggshell thinning and shells breaking. This was because DDE reduces the amount of calcium in the shell.

- The **lithosphere** is dominated by a layer of soil of varying depth and characteristics, produced by weathering of rocks and deposited by erosion. Because soils are solids, materials move much more slowly and over shorter distances than in air and water, making local movements more significant than global ones. Soil–groundwater movements are important for the transport of soluble materials into, and the leaching of materials out of, the soil layers. The type of soil is determined by

  - the underlying rock,
  - climate,
  - topography,
  - the biota and
  - the time over which the soils have been developing.

  Regional and local deficiencies in nutrients strongly influence terrestrial ecosystem processes.

- The **atmosphere** is the thin layer of gas surrounding the earth. It is divided into horizontal layers (Figure 3.11). The most important is the lower layer or troposphere which contains some 80 per cent of the mass of the atmosphere and which extends to a height of between 10 and 17 km. The stratosphere above the

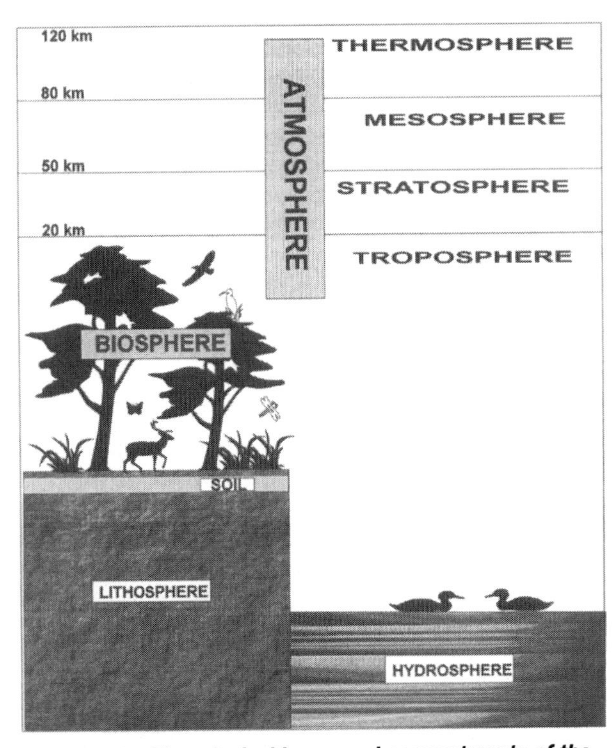

Figure 3.11 ***The principal layers and compartments of the Earth***

troposphere, is important because it contains the ozone layer that provides protection from most of the biologically damaging wavelengths of ultraviolet light. The atmosphere contains the largest pool of nitrogen and large supplies of oxygen. Although present in relatively small quantities, carbon dioxide is the main source of carbon for terrestrial primary producers. It also contains varying amounts of the water vapour so important to terrestrial forms. The atmosphere is a transport medium for many gases and fine particulate materials and exerts an enormous impact through its role in determining climate.

- The **biosphere** comprises all of the living components through which materials will circulate subject to the metabolic activities of organisms and their ecological interactions.

3.3.1 The hydrological (water) cycle

The cycling of very large quantities of water through the hydrosphere, atmosphere and lithosphere forms the hydrological cycle (Figure 3.12). Clouds and water vapour redistribute the water in the atmosphere after it evaporates from the surface of the Earth. Most comes from surface evaporation of the oceans but large amounts of water are also vaporised during *evapotranspiration*, the evaporative loss of water from the soil and from leaves. Evaporation from leaves drives plant circulation and cooling (**transpiration**). The energy for these processes comes directly or indirectly from solar radiation. Water returning to the surface as rain and snow is either taken up by plants or animals, percolates through the soil to the groundwater layer or runs off into streams, rivers and lakes. Eventually, all this water re-enters either the sea or the atmosphere and so the cycle continues.

The water cycle is a major factor modifying regional climates and is important because of its ability to move soluble substances such as nutrients and pollutants around within an ecosystem. Water is also important because it is essential to all forms of life.

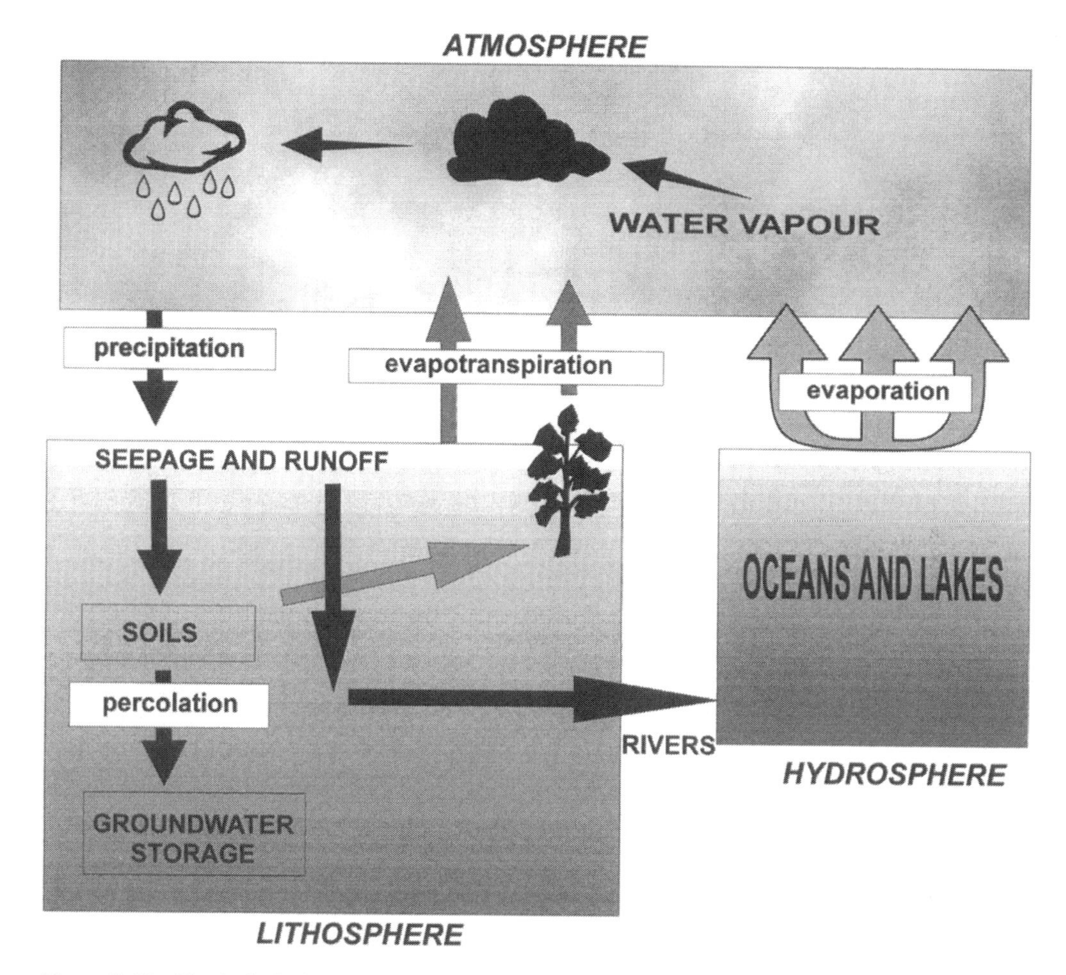

Figure 3.12 ***The hydrological cycle***

3.3.2 The carbon cycle

All life is based upon carbon chemistry. The gaseous carbon dioxide in the atmosphere (about 0.03 per cent) and non-gaseous bicarbonate and carbonate ions dissolved in water form major reservoirs of inorganic carbon, which is the origin of almost all biologically derived organic carbon. The abundant inorganic carbon bound into many sedimentary rocks interacts only slowly with these reservoirs through erosion and the weathering of carbon-containing rocks by rainfall. During the carbon cycle (Figure 3.13), biological processes redistribute carbon between the environmental compartments, the initial capture being through photosynthesis and the ultimate return being through respiration at all trophic stages, including decomposition. The organic carbon locked up in fossil fuels (oil, coal and gas) and wood is now being burnt and returned to the atmosphere as carbon dioxide at ever-increasing rates. This imbalance is causing a rise in atmospheric carbon dioxide levels (Figure 3.14), giving rise to speculation about global warming since carbon dioxide is a greenhouse gas.

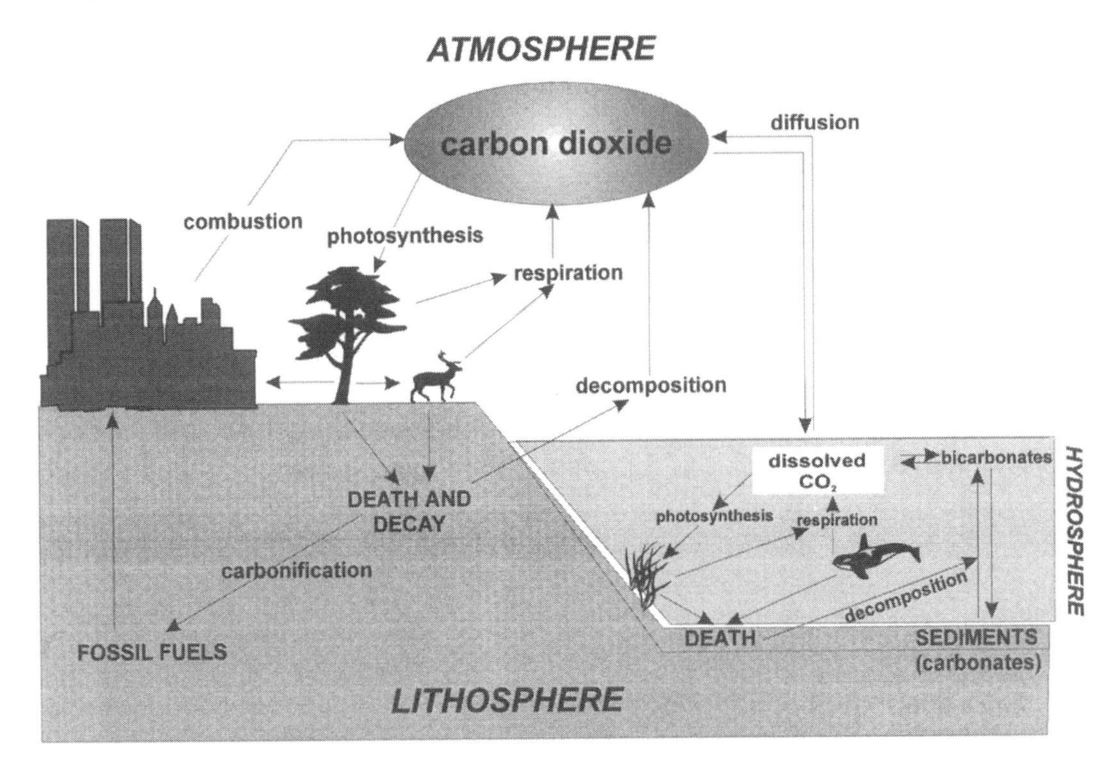

Figure 3.13 ***The carbon cycle***

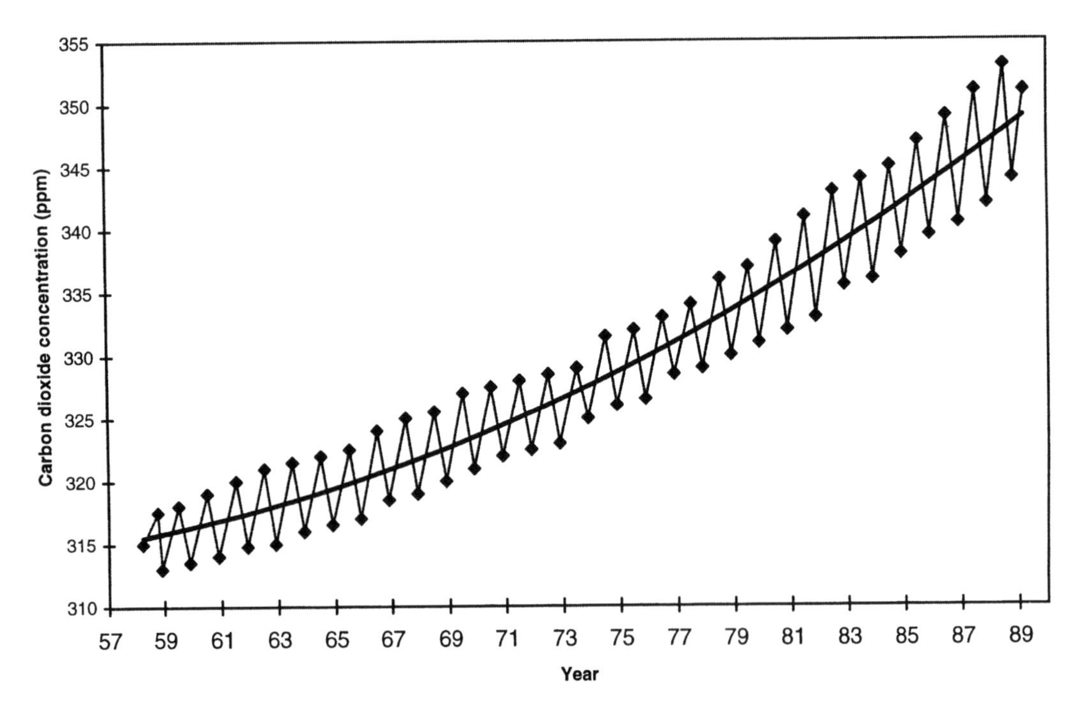

Figure 3.14 ***Changes in the mean monthly maxima and minima concentrations of carbon dioxide in the atmosphere measured at Mauna Loa, Hawaii. The steady increase in carbon dioxide is clear***

3.3.3 The oxygen cycle

Atmospheric oxygen, so vital to today's dominantly *aerobic* biota, comes from two main sources. Most comes from photosynthesis but small amounts also come from the dissociation of water molecules in the upper atmosphere. Today's atmospheric oxygen concentration remains at a fairly constant 20 per cent although this figure has varied during the Earth's history. Life is thought to have originated in an oxygen-free atmosphere when any oxygen present was chemically bound into inorganic compounds. The production of free oxygen depended upon the evolution of photosynthesis, an event which truly changed the course of Earth's evolution.

Atmospheric levels of oxygen are a balance between oxygen utilisation (mainly respiration) and oxygen production (mainly photosynthesis). The world's major forests and aquatic phytoplankton, therefore, are vitally important since they are both the main producers of oxygen and the absorbers of carbon dioxide through photosynthetic activity.

Oxygen in the outer atmosphere occurs as ozone (O_3). A vital layer of ozone encircles the Earth and shields the surface from the mutagenic and carcinogenic effects of ultraviolet light (UV). Protection from UV results from its interaction with ozone, the UV dissipating as heat thus:

$$UV + O_3 \rightarrow O + O_2 \rightarrow \text{reforms} \rightarrow O_3 + \text{heat}$$

The destruction of the ozone layer by pollutants such as chlorofluorocarbons (CFCs) and oxides of nitrogen is of considerable significance for the planet's life forms. Its effects on the human populations affected by the ozone hole in the southern hemisphere are already measurable as increases in human medical conditions such as melanomas (skin cancer) and cataracts.

3.3.4 Nutrient cycles

The fundamental cycles outlined for water and carbon are supported by cycles of the nutrient materials required by organisms, the main ones being nitrogen, phosphorus and sulphur. Note particularly the critical role of decomposers in all of these cycles.

3.3.4.1 The nitrogen cycle

Nitrogen is an abundant, chemically inert gas making up almost 80 per cent of the Earth's atmosphere. It is a vital component of many biomolecules such as proteins and nucleic acids (Section 1.1.3). However, the gaseous molecule is very stable and has to be transformed before it can be used by most organisms. Only a few species of bacteria and cyanobacteria can convert nitrogen gas to biologically useful forms (nitrogen-fixing organisms), a process requiring large amounts of energy. Thus, despite its atmospheric abundance, biologically useful nitrogen is frequently in short supply and often may be the limiting factor in an ecosystem. Once incorporated into

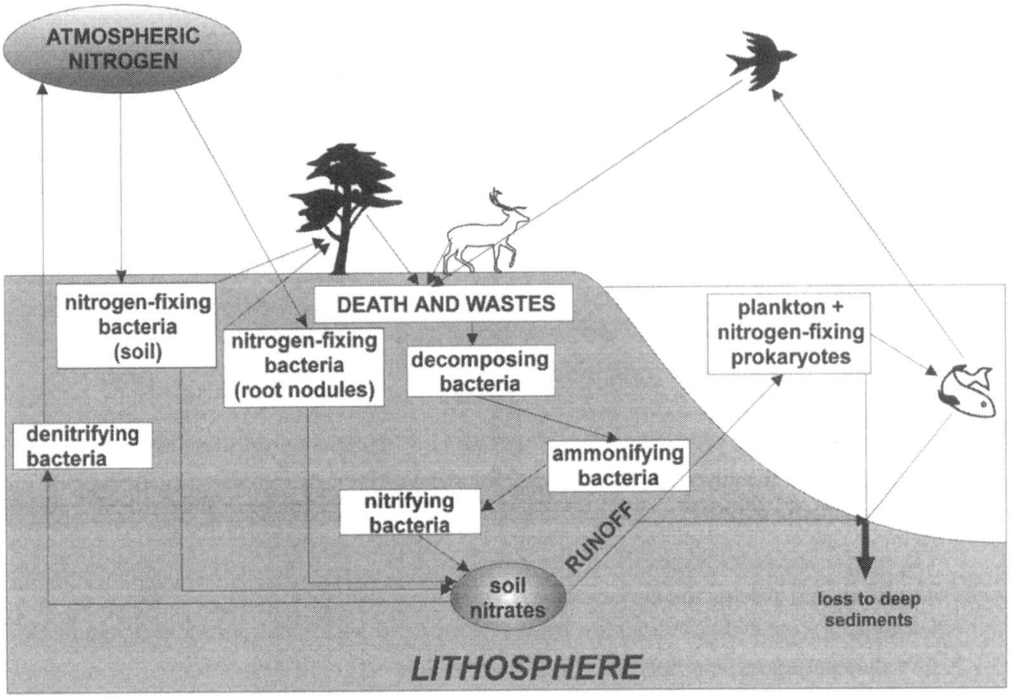

Figure 3.15 ***The nitrogen cycle***

biological materials, its recycling is dependent upon microbial decomposition processes. There are five main steps in the cycling of nitrogen (Figure 3.15):

1 **Biological nitrogen fixation**: the conversion of gaseous nitrogen to ammonia using an enzyme called nitrogenase that only works in the absence of oxygen and requires large amounts of energy. This capability occurs only in certain soil and aquatic bacteria and cyanobacteria (Box 3.8).

2 **Nitrification**: the conversion of ammonia to nitrate by soil bacteria. First *Nitrosomonas* and *Nitrococcus* species convert ammonia to nitrite and then *Nitrobacter* oxidises nitrite to nitrate, releasing energy in the process.

3 **Assimilation**: the uptake by primary producers of nitrate and/or ammonia derived from steps one or two and its incorporation into plant proteins and nucleic acids. These materials then pass along the food chain by digestion and further assimilation at each trophic level.

4 **Ammonification**: the decomposition of the nitrogen-containing waste products of organisms to ammonia by ammonifying bacteria in soil and aquatic environments. This ammonia recycles and is available for nitrification and assimilation.

5 **Denitrification**: the reduction of nitrate to unusable gaseous nitrogen, a steady drain from the biological component of the ecosystem. Anaerobic denitrifying bacteria carry out this process that is not extensive but does represent a continuous, significant loss of biologically active nitrogen (Box 3.9).

The nitrogen cycle is highly dependent on these bacteria and without them, life as we know it could not exist.

Box 3.8

Nitrogen fixation and root nodules

For many plants, scarcity of usable nitrogen compounds in the environment is one of the main growth-limiting factors. Gaseous nitrogen is abundant (78 per cent of atmosphere) but unavailable. Although conversion of gaseous nitrogen to a usable form is carried out by a few groups of prokaryotic (kingdom Eubacteria) organisms, it requires a great deal of energy to do so. This process of nitrogen fixation reduces gaseous nitrogen to ammonia using an enzyme called *nitrogenase* and frequently the nitrogen fixers live in intimate association with a specific eukaryotic organism. Their role in the biosphere is just as important as that of the photosynthetic autotrophs although their biomass is relatively small. They fix some 90 million tons of gaseous nitrogen per year, by far the most significant source of world nitrogen fixation. Various photosynthetic bacteria, including Cyanobacteria, are the main nitrogen fixers. On land, free-living soil bacteria make only a relatively small contribution to terrestrial nitrogen-fixation, most of it coming from the bacteria associated with the formation of root nodules in certain plants.

Some bacteria belonging to the genus *Rhizobium* live in close association (mutualistic symbiosis) with the roots of leguminous seed plants such as peas, soybeans, alfalfa and a variety of tropical shrubs and trees. Filamentous bacteria called Actinomycetes operate similarly in root nodules of shrubs such as alder. These nodule bacteria release up to 90 per cent of the nitrogen they fix to the plant and also excrete some amino acids into the soil, making some nitrogen available to other organisms. By this means, nitrogen-poor soils can both be cultivated and improved and for this reason, crop rotations usually involve the periodic growing of leguminous plants to help improve the soil.

The formation of nodules results from free-living *Rhizobium* bacteria interacting with the growth of root hairs. A nodule forms as the plant's reaction to the initial infection by the bacterium and the bacteria then multiply, rapidly filling the host cells with bacteria (**bacteroids**). The plants supply the bacteroids with sugars from the legume root cells and the bacteroids use these sugars to provide the energy for nitrogen fixation, providing a sophisticated symbiotic relationship. Interestingly, the critical bacterial enzymes are very sensitive to the presence of oxygen, being poisoned by even traces of free oxygen. However, it is the plant, not the bacteroid, which produces *leghemoglobin*, a compound that traps any oxygen present, thus protecting the bacterial enzymes from oxygen poisoning.

3.3.4.2 The phosphorus cycle

Phosphorus is another element essential to life through its use in phospholipids, nucleic acids and the energy-transporting ATP (see Section 1.1.2.3). Since it does not exist in a gaseous form, the atmosphere plays only a minor role in this cycle (Figure 3.16), for which the primary reservoir is sedimentary rocks. These erode, slowly releasing inorganic phosphate ions, which are biologically useful. Some phosphate runs directly into the sea via rivers and streams while much is absorbed by plant roots and thus enters the terrestrial food chain.

The aquatic and terrestrial cycles operate similarly but largely separately. Phosphates are eventually deposited on the sea floor and locked up as sedimentary

Box 3.9

Nutrient pollution: too much of a good thing

To increase and sustain yields of crops and to maintain attractive gardens and amenities, chemical fertilisers are widely used. In fact, some 25 per cent of the world's crop yield is attributed to their use. Chemical fertilisers are important because they replace soil macronutrients and micronutrients removed during the crop's growth and harvesting. Although they replace the inorganic nutrients, most chemical fertilisers do not replace the soil organic matter that is important for the maintenance of proper soil chemistry and the retention of the inorganic nutrients. Total dependency on inorganic fertilisers alone can result in detrimental changes in the physical, chemical and biological properties of the soil.

Water seeping through the soil leaches nutrients such as nitrates and phosphates and carries them away into streams and lakes. This loss is exacerbated when fertilisers are applied at either the wrong time of year, just before a period of heavy rain or in excessive quantities so that plants are unable to use the nutrients before they leach. It is estimated that up to a quarter of some fertiliser applications is lost into adjacent water bodies, increasing the nutrient status of the water bodies they enter. This causes increased primary productivity in these systems that are normally limited by the natural levels of (particularly) nitrates and phosphates. Consequently, the easily grazed spherical green algae are often replaced by more filamentous and often toxic Cyanobacteria, producing changes in secondary productivity at higher trophic levels. Secondary production may also increase as a result of the extra primary production or it may be markedly inhibited as a result of the secondary consequences of the increased plant production. Such effects include the development of toxic algal blooms, deoxygenation of the water column due to the increased respiration or decomposition, or physical clogging of delicate respiratory structures.

This nutrient enrichment of natural water bodies as a result of pollution is *eutrophication* and is most conspicuous for phosphorus, which is the most commonly limiting macronutrient. This problem is often exaggerated by the major nutrient enrichment which derives from the disposal of sewage containing phosphorus-rich detergents, human wastes and/or animal wastes. Although most emphasis is given to nutrient enrichment of freshwater lakes such as Lake Erie in America, eutrophication of shallow, coastal seas now occurs. The southern North Sea is subject to significantly increased nutrient loading from the major European rivers and adjacent landmass and some increases in toxic marine algal blooms are attributed to this eutrophication.

rocks for millions of years until raised to the surface again by geological activity. Phosphorus cycles much more rapidly through marine organisms than through terrestrial ones and most phosphate cycles within the biological components of an ecosystem. Autotrophs such as plants take it up and it passes along the food chain, eventually recycling once decomposers reconvert it to phosphate ions. It then enters the pool of ions contained in either the soil or the water column. On average, phosphorus atoms cycle about fifty times between surface waters and marine sediments before becoming locked up in those sediments. While the transfer of terrestrial phosphorus to the sea by drainage is continuous, transfer from the aquatic system to the land is relatively limited, mainly taking the form of sea bird defecation (guano) on land.

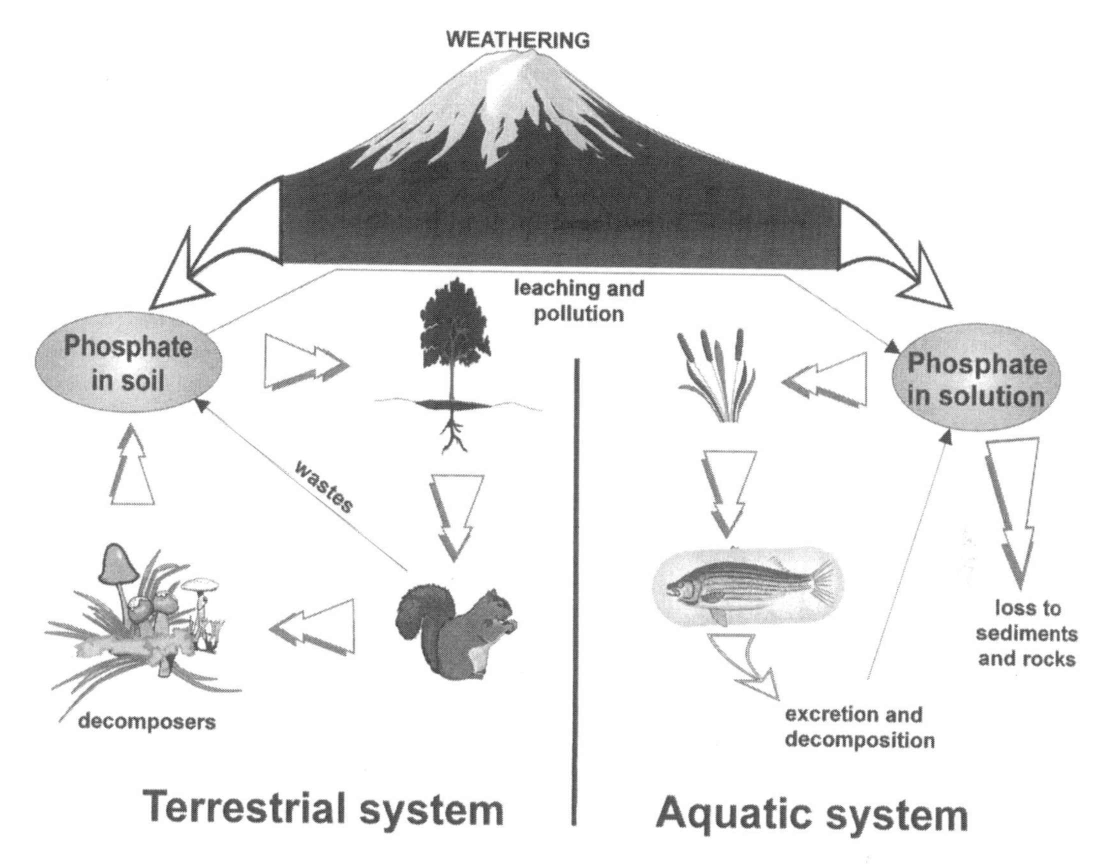

Figure 3.16 ***The phosphorus cycle***

Recently, anthropogenic sources of phosphorus such as sewage and fertilisers have disrupted the natural phosphorus cycle (cultural eutrophication). This has resulted in significant changes in the production and species content of aquatic systems in particular.

3.3.4.3. The sulphur cycle

Unlike nitrogen and phosphorus, sulphur is rarely a limiting factor for organisms although much of it is tied up in rocks and minerals. Figure 3.17 shows that sulphur enters the atmosphere from several natural sources. The breakdown of organic matter by microbial anaerobic decomposers releases hydrogen sulphide (H_2S), a poisonous gas smelling of rotten eggs. Volcanic activity is a major natural source of hydrogen sulphide and also produces sulphur dioxide (SO_2). However, much of the sulphur release comes from human activities such as the burning of fossil fuels and smelting of metal sulphides.

The impact of sulphur comes mainly from the reaction of sulphur dioxide with atmospheric oxygen and water to form tiny droplets of sulphuric acid (H_2SO_4) that fall to Earth as acid rain. This damages trees directly and freshwater life either directly

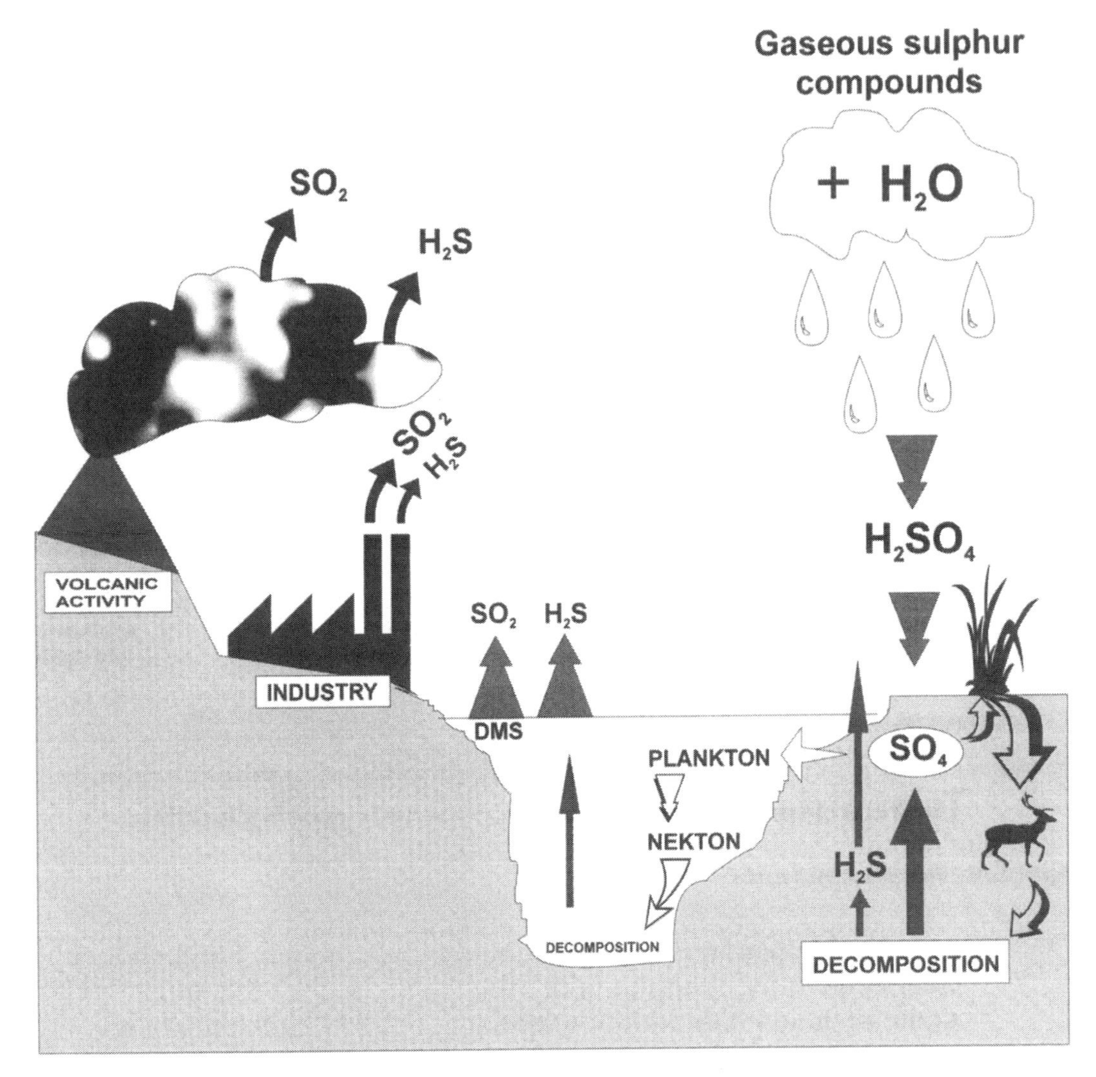

Figure 3.17 ***The sulphur cycle***

through acidification or indirectly through an increase in the leaching of toxic ions such as aluminium from watershed soils. Another source of atmospheric sulphuric acid is from the production of dimethylsulphide (DMS) by planktonic marine algae.

Sulphur plays an important role in the control of global climate. Now that anthropogenic sulphur production from the burning of fossil fuels, etc., is equivalent to natural sulphur fluxes, the enhanced quantities of sulphur in the atmosphere are potentially influencing climates and contributing to acid rain pollution.

Summary points

- The laws of thermodynamics control the transfer of energy through the ecosystem. The key feature is the loss of energy that occurs at every transformation within an ecosystem.

- Energy flow through an ecosystem is a one-way process unlike materials which are recycled. Thus light energy passes only once through any food chain and is eventually lost to the system. Chemicals are recycled within the environment as biogeochemical cycles.
- Energy transfer is subject to an array of variables and a variety of measures of efficiency have been determined. Systems differ in their energetic adaptations and properties.
- The concepts of trophic (feeding) levels, food chains and food webs are important for the understanding of energy transfer in any ecosystem because they identify the ways in which both energy and materials pass through the system.
- Feeding relationships are fundamental to the organisation and function of biological systems. Food chains are simple, comparatively rare and pose the risks of narrow specialisation. Food webs may be very complex, are widespread and have the safety factor offered by substitutable food resources.
- Every system has three main components: the autotrophic producers, the secondary producers (including the herbivores and carnivores) and the decomposers. They are generally equally important but in some habitats, one or two components may dominate.
- The world comprises four main compartments: the hydrosphere (water), the lithosphere (land), the atmosphere and the biosphere. Materials are distributed and cycle between these compartments.
- Chemical materials cycle within biogeochemical cycles, the main ones being the water cycle, carbon cycle, oxygen cycle and the nutrient (particularly nitrogen, phosphorus and sulphur) cycles. Many of these cycles are being disrupted to varying degrees by human activities, both industrial and agricultural.

Discussion / Further study

1. Why are most ecological pyramid structures misleading or incomplete? Try to construct a pyramid for a habitat of your choice but include *all* components!
2. What are the advantages and disadvantages of using commercial chemical fertilisers? Why should both inorganic and organic fertilisers be used?
3. Why is a vegetarian diet more energy-efficient from a world food viewpoint?
4. Why are phosphates cycled more rapidly through aquatic habitats?
5. How does acid rain cause environmental damage?
6. How is ozone produced in the atmosphere and how is the outer atmospheric ozone layer being affected by human activities?
7. The problems associated with the use of DDT have been outlined. Study the history of pollution by mercury, cadmium and PCBs to derive the significant environmental messages that derive from these experiences.

Further reading

Environmental Ecology: The effects of pollution, disturbance and other stresses, 2nd edition. Bill Friedman. 1993. Academic Press, San Diego.
A good survey of the problems of pollution, dominated by American examples.

'Red tides.' D.M. Anderson. 1994. *Scientific American* 271 (2), 52–58.
A good account of marine algal blooms and their significance.

'The toxins of Cyanobacteria.' W.W. Carmichael. January 1994. *Scientific American* 270, 64–72.
A closer look at their poisons and their potential health effects.

Understanding our Environment: An introduction to environmental chemistry and pollution. R.M. Harrison. 1992. Royal Society of Chemistry, Cambridge.
A valuable basic text for those not up to speed on their chemistry.

Principles of Geochemistry. B. Mason, and C.B. Moore. 1982. John Wiley and Sons, New York.
Another useful book for those wanting to know about biogeochemical cycling in more detail.

Environmental Science. A.R.W. Jackson and J.M. Jackson. 1996. Longman Group Ltd, Harlow, England.
A very good and easily readable overview of environmental science topics.

References

Carson, Rachel. 1962. *Silent Spring*. Penguin Books, London.

Colborn, T., Myers, J.P. and Dumanoski, D.L. 1996. *Our Stolen Future*. Brown and Co., Boston.

Elton, C. 1927 *Animal Ecology*. Sidgewick and Jackson, London.

Kurihara, Y. and Kikkawa, J. 1986. 'Trophic relations of decomposers', in J. Kikkawa, and D.J. Anderson (eds) *Community Ecology: Pattern and process*, Blackwell Scientific Publications, Melbourne.

Miller, G.T.J. 1994. *Living in the environment: Principles, connections, and solutions*, 4th edition. Wadsworth Publishing Co., Belmont, CA.

Teal, J.M. 1962. 'Energy flow in the salt marsh ecosystem of Georgia', *Ecology* 43, 614–662.

Conditions and resources: major determinants of ecology

Key concepts

- **Conditions are environmental factors which influence organisms but are never consumed or used up by an organism – they require both adaptations to and tolerance of the conditions by organisms.**
- **Resources are often limited environmental factors which are consumed or used up by organisms and which are the source of competition between organisms.**
- **Each species occupies an ecological niche, a multidimensional (conditions and resources) concept of the total requirements of a species.**
- **Evolutionary adaptations occur in response to both conditions and resources and an adaptation by one species usually leads to a new responsive adaptation in associated species.**

The distribution and abundance of a species is influenced by a variety of factors. These include evolutionary history, rates of birth, death and migration (see Chapters 5 and 6) and the nature of intraspecific (within a species) and interspecific (between species) interactions. However, the dominant factors for most species can be classified as either conditions or resources; some can be considered as both.

A *condition* is an abiotic environmental variable which fluctuates in both space and time and to which organisms vary in their response: examples include temperature, moisture, pH and salinity. Although the presence of other organisms may modify a condition, it is never consumed or used up by the organism. It is never, therefore, the source of competition although it is frequently the cause of adaptive evolution. This clearly distinguishes it from a *resource* such as a nutrient, food item or space, which organisms consume or use up and which is often the source of intense intraspecific and/or interspecific competition (see Section 6.2.6). The responses of organisms to conditions and resources led to the concept of the ecological niche as proposed by Hutchinson (1957). A niche is best described as a multidimensional (conditions and resources) space (hypervolume) which can support a viable population (Box 4.1).

Before we consider conditions and resources individually, two 'laws' which markedly influence the significance of variations in these parameters need consideration.

Liebig's Law of the Minimum (Liebig 1840) states that

> the total yield or biomass of any organism will be determined by the nutrient present in the lowest (minimum) concentration in relation to the requirements of that organism.

Box 4.1

The ecological niche: a multidimensional hypervolume called home

The term 'niche' (Elton 1927) is used in a variety of contexts such as:

- *spatial*, which includes habitat, distribution and physical location
- *functional*, which includes trophic levels and other food web relationships
- *behavioural*, which refers to mode of life such as predator, competitor, etc.
- *abstract*, as a multidimensional space composed of interactions with diverse environmental factors, both abiotic and biotic.

In all natural communities, successful functioning depends upon the operation and integration of different 'professions', each of which occupies this multidimensional space called a niche. A niche represents the sum of

- an organism's adaptations,
- its use of resources and
- the lifestyle to which it is fitted.

Each niche is occupied by a different species, although ecologically equivalent species may occupy comparable niches in different or similar ecosystems in other parts of the world.

Some niches occur in every major community because they perform an indispensable function such as primary production. Others are widespread but restricted by certain physical conditions. Thus *bioturbation* processes (the vital biological aeration and mixing of soils and sediments) are typically carried out by earthworms in moist soils but not in dry soils because earthworms cannot live in dry soils. Evolution can result in the development of highly specialised niches such as those occupied by organisms that provide a valet service for larger animals. Certain birds carry out this function in terrestrial regions and on various, host species while coral reefs have their own equivalents such as cleaner wrasse and shrimps.

Niches may be divided into

- **fundamental niches**, the theoretical multidimensional niche that would be occupied in the absence of other species and
- **realised niches**, which are the actual niches occupied in a community where there are interactions and sharing of resources with other competitive species.

No two species can occupy exactly the same multidimensional niche in the same community for any length of time because *competitive exclusion* will occur (see Section 6.2.6), one species eliminating its competitor. However, fundamental niches can overlap, leading to species coexisting and competing. It is this competition between different species that defines a species' realised niche.

Although this refers specifically to the availability of nutrient resources, it draws attention to the fact that the success or failure of an organism is determined by that factor or factors (condition or resource) which is limiting.

This led to *Shelford's Law of Tolerance* (1913), which states that:

> for an organism to succeed in a given environment, each of the conditions must remain within the tolerance range of that organism and that if any condition exceeds the maximum or minimum tolerance of that organism, the organism will fail to thrive and will be eliminated.

Thus, the distribution and abundance of a species is controlled by those conditions or resources which limit its success and these may vary with the habitat in which the species finds itself.

4.1 Conditions

The main conditions influencing distribution and abundance of species or individuals include:

- temperature
- moisture and relative humidity
- pH of soil and water
- salinity
- current flow
- soil structure and substratum
- pollutants.

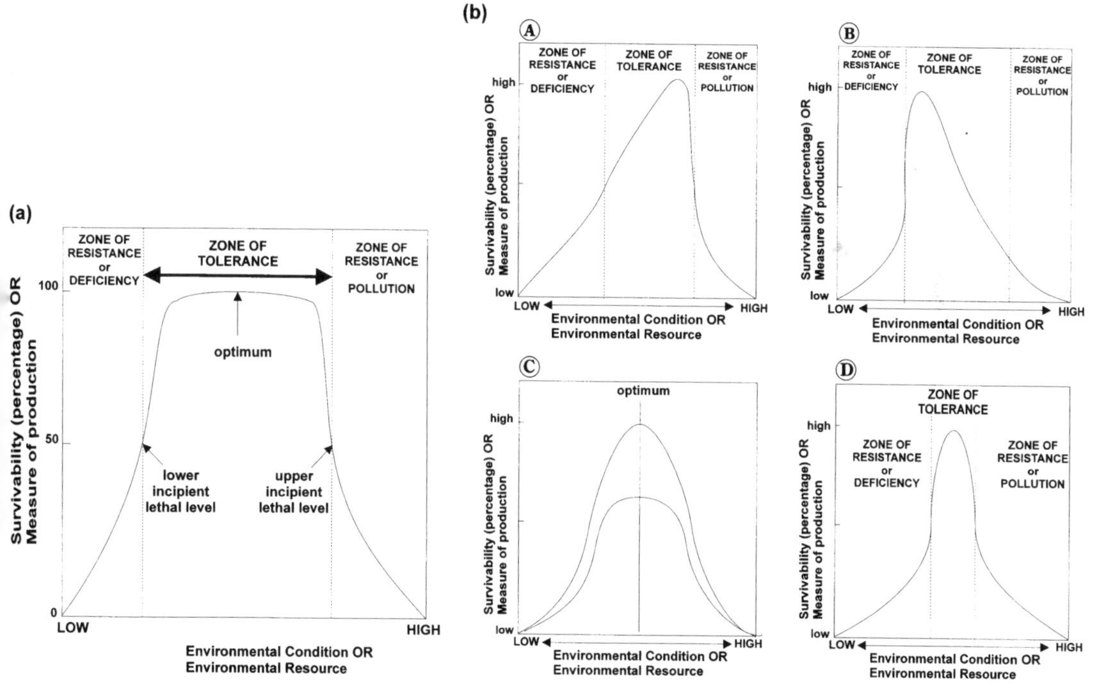

Figure 4.1 ***Responses of organisms to stressors. (a) A generalised response curve to show the main terminology associated with a response curve. (b) Examples of variations in skewness and kurtosis which may occur in response curves***

The response of any species or individual to each condition can be described by a *response curve* (Figure 4.1a), allowing the identification of optima and ranges for each combination of condition and species/individual. However, this will depend upon the parameter used to measure performance and the varieties are very difficult to measure in practice. The effects of a range of conditions on different measured parameters will vary also. The exact shape of the response curve will also vary with each condition and between individuals and species: thus it may be symmetrical or skewed, broad or narrow (Figure 4.1b).

Consider now each of the major conditions and their significance.

4.1.1 Temperature

Environmental temperature variations have considerable potential for the determination of the distribution and abundance of organisms. There are many types and sources of variation, the main ones being:

- **Latitudinal variation**: The world is divisible into generalised temperature zones (Figure 4.2) which are related to the broad distribution of the world's major biomes (see Chapter 3). However, such broad categorisations are of little real value since they obscure the highly specific nature of temperature variations on daily or seasonal scales where the Law of Limiting Factors becomes very significant.
- **Seasonal variation**: As a consequence of the tilt of the Earth's axis and its elliptical rotation around the sun, there are seasonal variations in temperature which are related to latitude. There are only small variations in temperature near the equator while the middle and higher latitudes may show very dramatic fluctuations, the extremes of which exert a major biological influence.
- **Altitudinal variation**: There is a decrease of 1°C for every 100 m increase in altitude in dry air and 0.6°C in moist air. These changes are superimposed upon the large-scale geographical trends and may provide refuges for cold-adapted species in the warmer areas of the Earth. Thus northern, cold-adapted plant species may extend south by living high in mountain ranges.
- **Continentality**: Because water bodies do not warm or cool as rapidly as land masses, proximity to the sea has a moderating effect on temperature variation, giving rise to the 'maritime' climate of coastal regions and islands. Similar effects occur within land masses where dry, bare desert areas are subject to larger daily fluctuations than are, for example, forest areas.
- **Microclimatic variation**: Temperature may vary on a very local scale, having very significant effects on the ecology of organisms. Some examples include the collection of cold, dense air in a valley, frost hollows, local heating by insolation and the effects of dense plant canopies in modifying air temperature.
- **Depth variation**: Increasing depth in soil or water reduces and delays temperature fluctuations taking place at the surface. Thus a metre or so below a soil or sediment surface, diurnal variations of temperature may be absent and, deeper still, even annual variations may be absent. In water bodies, temperature fluctuation diminishes markedly with depth so that below about 1000 m, all oceans are uniformly cold and have no seasonal or diurnal temperature variation.

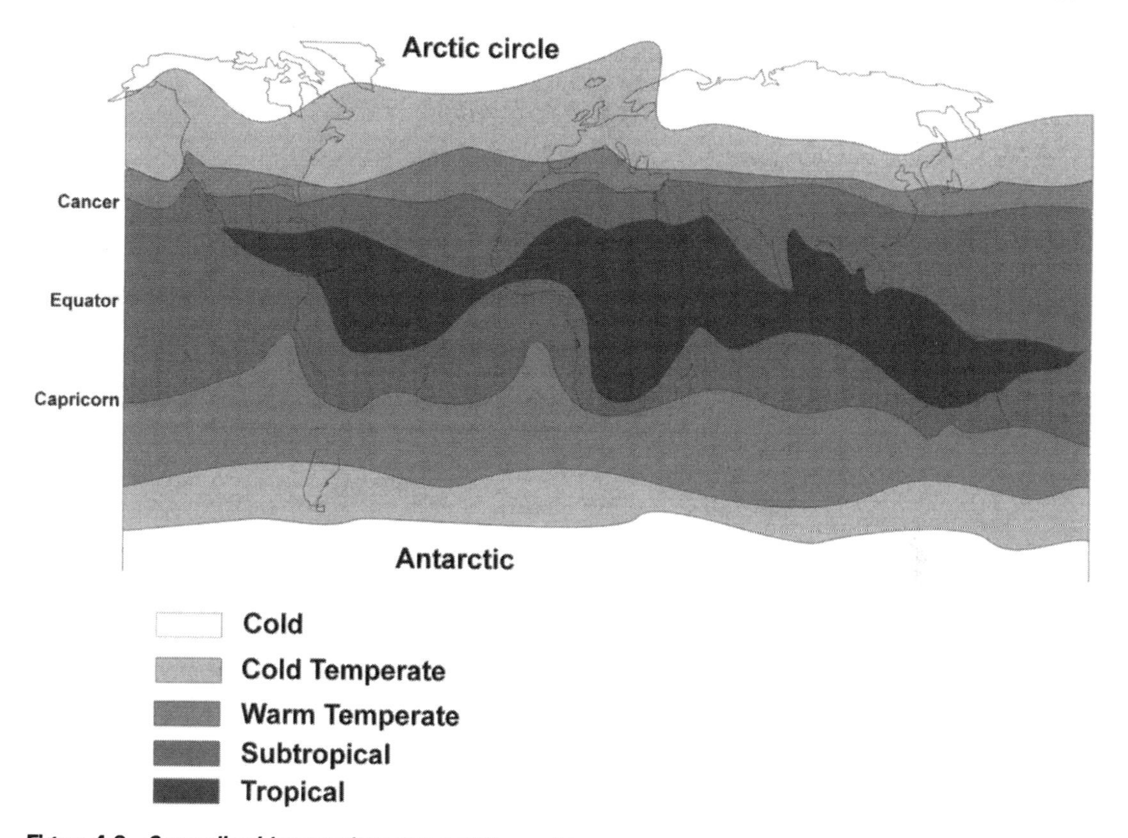

Figure 4.2 ***Generalised temperature zones of the world***

4.1.2 Moisture

The importance of water has already been discussed (Section 3.3.1) but for aquatic organisms, unless there are osmotic problems, water is not a limiting factor. However, the evolution onto land and adoption of a terrestrial mode of life posed very significant problems. Air usually has a lower water concentration than the organisms and there was a serious risk of dehydration through evaporation and loss of water during excretion. Successful colonisation of land, therefore, required development of mechanisms for the control of water loss.

A key condition for land organisms, therefore, is the relative humidity of the atmosphere. The greater the humidity, the smaller is the differential between the atmosphere and the organism and the less the risk of desiccation. Humidity and availability of water in an environment are intimately linked: a desert has both a low humidity and a shortage of available water. Relative humidity is also intimately related to temperature since temperature and rate of evaporation are related. Other parameters which also affect rate of evaporation, such as wind, also interact significantly with humidity. Figure 4.3 summarises the interrelationships between moisture and other key parameters.

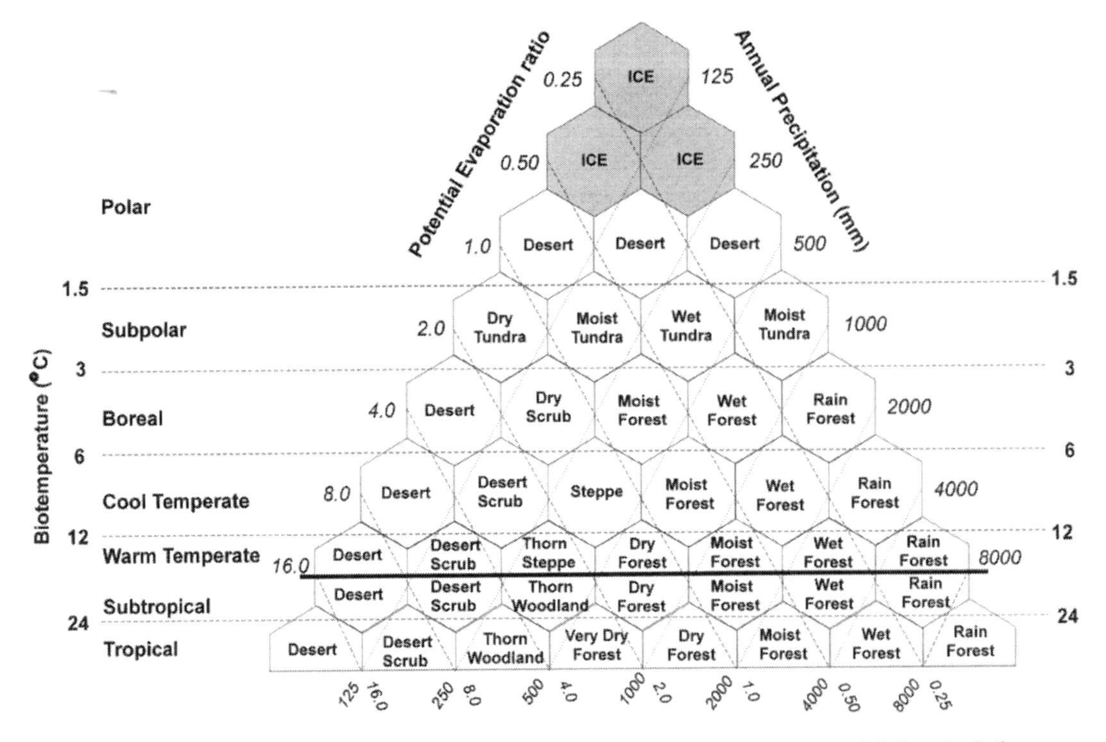

Figure 4.3 ***Holdridge's Life Zone classification scheme. This relates temperature, rainfall and relative humidity (evaporation ratio) to each other to provide a habitat classification scheme***

The differing abilities of organisms to control water loss have major significance for their distribution in habitats which themselves vary in their moisture content. Terrestrial plants differ from animals in two fundamental respects:

- Although aerial parts suffer evaporative loss as transpiration, their underground roots are usually in contact with soil from which water can be obtained.
- Water is both a condition and a resource for plants because it is required for photosynthesis and thus for plant nutrition.

However, relative humidity is a very important factor controlling the distribution of all organisms, including plants. Organisms with poor control of water loss such as earthworms and other soft-bodied animals must live in damp habitats. Organisms such as kangaroo rats and cacti have special water-control processes and can live in harsh desert environments (see Box 5.1).

4.1.3 pH

The pH of soil or water is a condition influencing the distribution and abundance of organisms. A pH outside a range of 3 to 9 is usually damaging to organisms, either directly or indirectly, because soil pH affects nutrient and toxin availability and concentration. The range of tolerance is even narrower if the criteria are the growth and reproduction of most plant species. In acid conditions, the low pH may result in high and toxic concentrations of aluminium ions (below about pH 4), manganese ions and iron ions, despite the latter two being essential plant nutrients under other conditions. In alkaline conditions, many essential plant nutrients become locked up as insoluble, and therefore unavailable, compounds.

For aquatic animals, acidic conditions usually result in reduced diversity because they:

- interfere with **osmoregulation** (the regulation of water balance in an organism), enzyme activity or function of respiratory surfaces;
- increase the concentrations of toxic ions in the water;
- decrease the range and type of food resources.

There are, therefore, parallels between the way pH affects the environment in both soil and water; both show direct and indirect effects. The indirect effects derive from interactions either between two conditions (e.g. pH and toxin concentration) or between a condition and a resource (e.g. pH and nutrients or food) (Box 4.2).

4.1.4 Salinity

Sea water contains abundant salts (Table 4.1), which results in many organisms being isotonic (having a similar concentration of salts in their body fluids) with the water, making water and salt control simple. However, some organisms, such as the bony fishes, are *hypotonic* (having a lower concentration of salts in their body fluids) to normal sea water and they risk the loss of water through *physiological desiccation*. This is effectively the same problem encountered by terrestrial species. For such aquatic organisms, the control of body fluids is an essential but energy-consuming process since they have to operate ‘pumps’ to maintain their body fluids against osmotic gradients. The salinity of an aquatic environment can, therefore, exert very significant effects on the distribution and abundance of species because of differences in the water and ionic-control abilities of different organisms. This is particularly true where the salinity is increased (salt pans) or decreased, either with short-term variation as in estuaries and rock pools or in more stable brackish-water lagoons. It can also be very significant in habitats bordering the sea (salt-marshes, mangrove swamps, etc.). Organisms vary markedly in their tolerance of salinity and salinity fluctuations. Salt-tolerant plants (*halophytes*) tolerate high salinity by concentrating electrolytes in the cell vacuoles while keeping the cell cytoplasm concentrations normal. Animals which can tolerate salinity fluctuations (termed **euryhaline**) do so by one of two methods (Figure 4.4):

- their cells may be tolerant of osmotic changes (**osmoconformers**) or
- they may physiologically regulate their body fluids to minimise change (**osmoregulators**).

Stenohaline species are intolerant of salinity variation and include organisms such as echinoderms, which are confined, therefore, to areas of stable and full salinity.

Box 4.2

Acid rain

Most of Europe and broad areas of North America and other industrialised regions are receiving precipitation that is between ten and a thousand times more acidic than is normal. However, this is not a new phenomenon and there is a natural component that is related to volcanic activity: studies of lake sediments have indicated that long before industrialisation, periods of acidic precipitation occurred in some areas. However, most of the current acidic rainfall derives from the interaction of the common industrially produced, gaseous sulphur oxides and nitrogen oxides with water vapour, producing an assortment of secondary pollutants. These form dilute solutions of nitric and sulphuric acids in the rain thus:

$$O_3 > O > OH + SO_2 + O_2 > H_2SO_4$$

ozone > oxygen atom > hydroxyl radical + sulphur dioxide and oxygen > sulphuric acid and

$$O_3 > O > OH + NO_2 + O_2 > HNO_3 \text{ (nitric acid).}$$

The result is rainfall with a pH as low as 3 compared with the unpolluted norm of 5.6. This slight acidity is normal and results from the solution of carbon dioxide in the water vapour to form carbonic acid.

The effect of acidification on populations of animals and plants is well established and it has become evident that this is the result of a variety of mechanisms of action. The pH of an environment affects the functioning of many aspects of biochemistry and deviations are very significant. Many freshwater systems have a natural pH of between 6 and 8 and organisms are adapted to this range. Acidification may result in direct mortality or it may affect organisms indirectly such as by interfering with reproduction. A very significant problem is the leaching of various heavy metals and aluminium from the soil as acid water percolates through it. Such elements are usually bound within insoluble mineral compounds and are not biologically active but acidification of the watershed washes these substances into freshwater systems. The results have included massive fish kills so that in Ontario, Canada, for example, about 1200 lakes are now devoid of life.

Trees also appear to be affected significantly and many forests in America and Europe are showing signs of canopy damage and symptoms of what is termed forest decline. This is characterised by the gradual deterioration and often death of the trees. However, the complexity of forest systems makes a definite cause–effect relationship difficult to prove despite the high correlation between acid rain and the damage observed so far.

4.1.5 Movement of air and water

The movement of gas or fluid media influences the distribution of many organisms. It may be used for dispersal but it may also affect distribution and survival. The movement of air may be very significant in the context of the ventilation of an organism or aeration of a microhabitat. It may also be an important desiccating agent, particularly in regions of almost constant windy conditions such as the upper regions of mountains.

Table 4.1 ***The composition of sea water***

	Ion	Percentage
Major constituents (>100 ppm)	Chloride (Cl^-)	55.04
	Sodium (Na^+)	30.61
	Sulphate (SO_4^{2-})	7.68
	Magnesium (Mg^{2+})	3.69
	Calcium (Ca^{2+})	1.16
	Potassium (K^+)	1.10
	Element	*ppm*
Minor constituents (1–100 ppm)	Bromine	65.0
	Carbon	28.0
	Strontium	8.0
	Boron	4.6
	Silicon	3.0
	Fluorine	1.0
Trace elements (<1 ppm)	e.g. Nitrogen, Phosphorus, Iodine, Iron, Zinc, Molybdenum	

The greater density and viscosity of water make its movement physically more significant than air but it obviously does not have any desiccating consequences. Shores are subject to wave action that dramatically affects the distribution and abundance of different species. Water is heavy and a breaking wave is potentially very damaging; to survive in areas exposed to wave action, organisms therefore require special adaptations (Box 4.3). Currents can be limiting also and the distribution of many species is determined by their degree of tolerance of current flows. Currents impose the need for morphological and behavioural adaptations to allow survival in the more extreme conditions (Box 4.4). They also provide a means of replenishing materials such as oxygen, nutrients and food vital to organisms such as plants and barnacles that cannot move to obtain such materials.

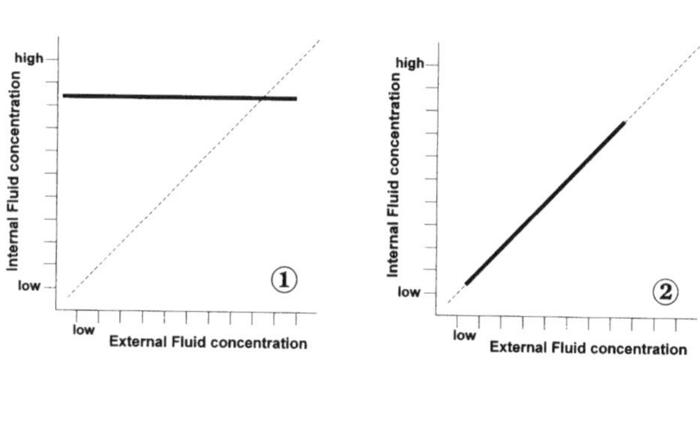

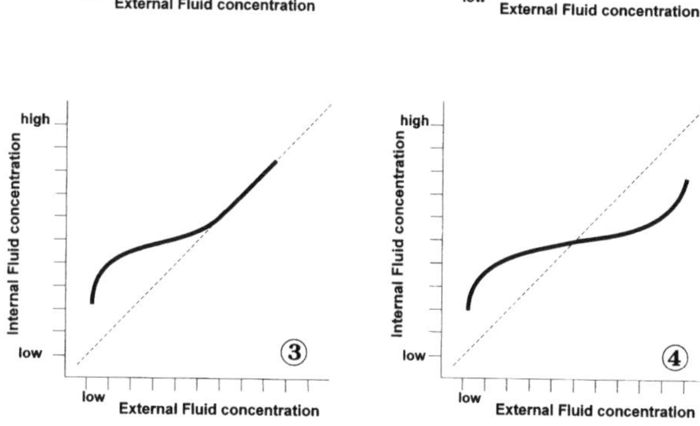

Figure 4.4 ***Four generalised types of osmoregulatory response. 1, Perfect osmoregulation. 2, Perfect osmoconformer. 3, Hyperosmoregulation in dilute media. 4, Both hyper- and hypo-osmoregulation. The dotted line represents the line of iso-osmicity***

4.1.6 Substrate type and structure

The physical and chemical nature of the substrate which the organism lives on or in can be of great significance in the determination of distribution and abundance. Rocky substrates vary in hardness, chemical composition and geological formation while particulate substrates such as sand, mud and soil vary in factors such as particle size, mineral content and organic content. All of these variables have consequences for water and nutrient or pollutant concentration and availability. Organisms living on solid substrates such as rocky shores have often evolved special mechanisms for attachment such as the holdfasts of large algae and the adhesive foot of the limpet (Box 4.3).

Particulate substrates, however, require different attachment strategies. Coarse sediments with low organic contents tend to be unstable and have a low water content (e.g. coarse sands in deserts or estuarine sandbanks). Finer sediments retain water through *capillary action* but this same process also tends to restrict circulation of the interstitial water (the water between the particles): this leads to anoxic conditions and nutrient limitation in the deeper sediment layers. This is a particular problem in aquatic sediments. Muds and silts contain the smallest particles and exhibit these problems to the maximum extent. Survival in such habitats usually requires a burrowing lifestyle with the maintenance of a direct connection to the overlying water to ventilate the burrow.

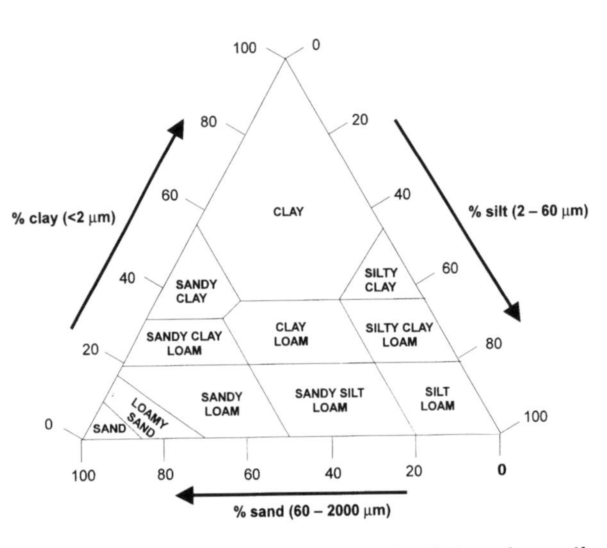

Figure 4.5 ***Very simple classification of soils based upon the relative abundance of three sizes of particle (after Chapman and Reiss 1992)***

Terrestrial soil structures are complex and variable and are major determinants of plant distributions. All soils comprise mineral particles of various sizes and chemical components derived from weathered or eroded rock together with organic matter in various stages of decomposition, mainly by fungi and bacteria. It is the differences in the relative proportions of these constituents that gives rise to the different soil types which influence the distribution and species composition of the contained communities. Soils are themselves affected by the plants that live in them. A simple classification of soil types based upon the relative abundance of different-sized particles is shown in Figure 4.5. However, because of the variety of soils found in different parts of the world, soils are also classified in a similar way to that used for organisms, i.e. there is a soil

Table 4.2 ***Soils of the world and their associated vegetation types***

Great Soil Group	*Vegetation type*
Laterite (latosol)	Wet tropical forest
Margalite (tropical black)	Seasonal tropical forest
Desert soil	Desert
Chernozem	Grasslands (steppe)
Chestnut soils	Grasslands (prairie)
Grey-brown podzol	Deciduous forest
Podzol	Boreal conifer forest
Peats	Mires/bogs
Tundra soil or well-drained podzol	Tundra

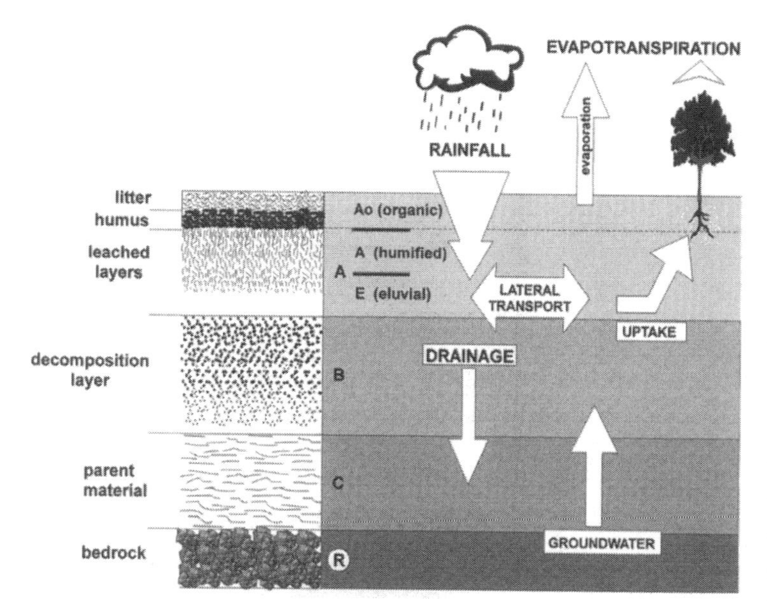

Figure 4.6 ***Generalised soil profile and classification of soil horizons. The diagram also shows the major water movements exerting an influence on soil conditions***

taxonomy. Most of the world's soil types are categorised into what are termed the Great Soil groups. These are usually associated with the major vegetation types (Table 4.2) and with climatic factors such as rainfall and annual temperature cycles.

Soils are typically composed of complex vertical layers or *horizons* that together form the soil profile. Figure 4.6 illustrates the key processes that take place in this stratified system and which result in the movement of materials, both useful and damaging, into and out of the system. Plants often have structural adaptations to the particular substratum on which they live. Thus, the pneumatophores of mangroves allow their roots to breathe in estuarine mud environments and the buttress structures of tall trees provide the additional support needed for living on the thin soils of tropical rainforests.

4.1.7. Contaminants and pollutants

The presence in the environment of persistent or non-persistent materials that may be toxic, causing lethal or sub-lethal effects, has led to the recognition of:

- *contaminants* – materials present at levels which do not cause a biological effect and which may derive from natural or anthropogenic sources
- *pollutants* – materials present at concentrations which exert a negative biological effect and which usually derive from anthropogenic sources.

Such materials may be *essential* or *non-essential*: organisms respond differently in each case (Figure 4.7). Different organisms show differing sensitivities to such materials, tolerant species having a variety of mechanisms providing them with protection from the polluting substances. The tolerance of seals to mercury provides a good example, being associated with increased levels of selenium in these animals. The development of tolerance is a crucial evolutionary process visible in the study of pollution ecology. It has far-reaching significance in the development of tolerance to pesticides by insects and to antibiotics by bacteria.

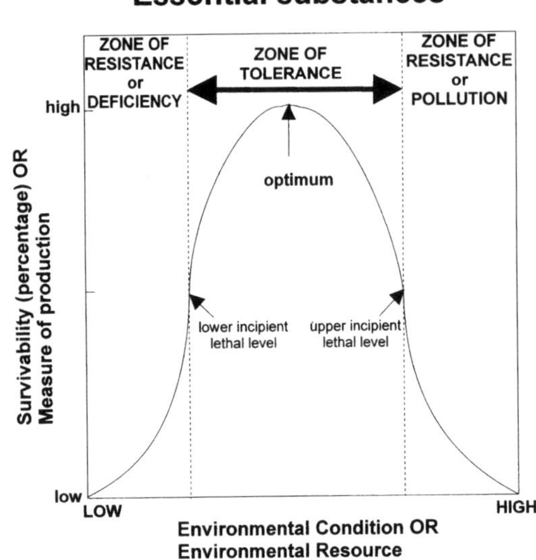

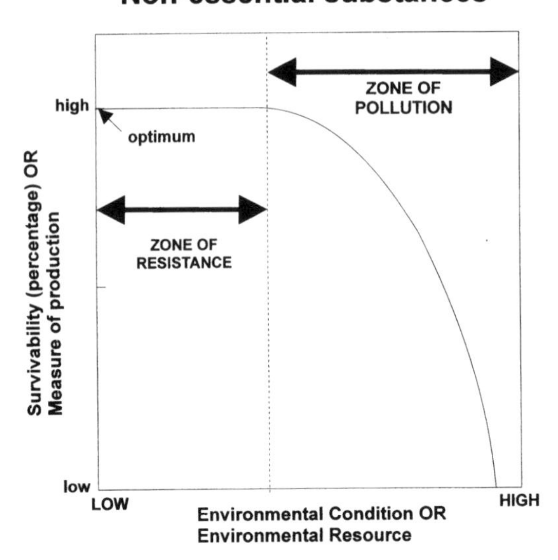

Figure 4.7 ***The two fundamentally different response curves of materials which can be considered essential and those which have no known biological function***

4.2 Resources

The resources of living organisms are mainly the materials from which they derive the substances necessary for the maintenance of their bodies, although space is also an important resource for most organisms. They are parameters that can be reduced (consumed) by the activity of organisms and they are, therefore, the source of competition between organisms (in contrast with conditions, for which there is no competition, only individual tolerances). Much of ecology is about the synthesis of organic resources by plants from inorganic resources using sunlight (energy resource) and the subsequent capture and reassembly of these resources at each successive stage of a food web through consumer interactions.

The main ecological resources are:

- solar radiation
- inorganic molecules, i.e. nutrients
- organisms as food
- space.

Resources may be categorised as essential or substitutable. *Essential resources* are not replaceable by an alternative and the Law of Limiting Factors applies (see Chapter 4 Introduction). Nutrients such as nitrates or phosphates are clearly not substitutable by alternatives. The obligate hosts in a parasite life cycle are similarly essential resources. The more specialised an organism becomes, the greater is the tendency for resources to become essential in character. Two resources are substitutable when one can replace the other, either partially or totally; they may not be equally as good as each other but they can both be utilised. Thus a seabird

Box 4.3

Wave action: a major problem for organisms

Water is a heavy medium and the physical impact of waves on shores is evident at many scales: coastal erosion is a reflection of the power of wave action. Variations in coastal topography and fetch (the distance over which wind blows across the water uninterrupted by a landmass) cause tremendous variation in the intensity (size and frequency) of wave impact (wave shock) from place to place along the shoreline. This occurs on geographical, local and microhabitat scales. Wave action also markedly affects the substrate type on the shore because of its capacity to suspend and transport particles: the stronger the wave action, the coarser the substrate type will be and vice versa. The degree of adaptation to the forces imposed by wave action markedly affects the distribution and abundance of shore organisms.

Some organisms cannot withstand any significant degree of wave shock and are confined to sheltered habitats where they may also require to be adapted to the presence of deposited particulate materials. However, to live in the more exposed habitats, organisms require specialised forms of protection:

1. a **means of strong attachment** to prevent them being washed away. Seaweeds have holdfasts while mussels use strong, proteinaceous byssus threads secreted from a gland in the foot. Barnacles actually glue themselves to the rock surface. Mobile species such as limpets and chitons use their foot as a powerful suction-cup and fish such as lumpsuckers also develop a suction device to resist wave action.
2. a **physically resistant body structure** such as the calcareous shells of barnacles and limpets (generally conical and low profile to aid resistance) or a strong, flexible structure as in the case of algae such as kelp. Some red algae have assumed an encrusting form that is very resistant to wave action.
3. a **behavioural adaptation** leading to the seeking of sheltered microhabitats such as cracks and crevices. Many intertidal animals move towards local shelter when wave action increases. Organisms such as dog-whelks that wish to feed on barnacles and mussels on exposed rock surfaces return to adjacent sheltered areas during periods of intense wave action. Species such as barnacles and mussels are often rugophilic (preferring to settle in cracks) in habit, these cracks and crevices providing better attachment and some degree of protection at the same time. Mobile species such as crabs frequently move to protection under stones. A gregarious habit can also afford local protection by clumping organisms together.

The consequence of wave action is the development of distinctive communities under particular conditions of exposure to or shelter from wave action. These patterns are well established for the rocky coasts of Britain (Lewis 1964).

---

might eat either sandeels or sprats, each being a substitutable resource for that bird. Resources are *perfectly substitutable* when either can wholly replace the other or *complementary* when consumption of two resources simultaneously results in less being required than when they are taken separately. *Antagonistic resources* are the opposite of complementary ones in that more are required when they are consumed together than separately. This is usually associated with the consumption of materials with toxic interactions.

These will now each be considered briefly.

Box 4.4

Adaptations to life in a flowing environment

Lotic (flowing) environments, like wave-splashed rocky shores, present organisms with a significant but variable problem as a result of the physical impact of flowing water. The stones and boulders, spaces between them and the interstitial spaces within gravels and sands provide streams with a complex but periodically unstable architecture because of the variations in water flow which occur. This architecture also provides diverse microhabitats for organisms. The unidirectional flow, together with the power of flowing water, presents organisms with considerable problems since, if they are dislodged, they will be washed downstream and may die as a result of encountering unsuitable conditions. Organisms have, therefore, evolved a variety of adaptations to prevent or counteract such a hazard.

1 Micro-organisms such as bacteria, fungi, algae and protozoa living on permanently wet rocks produce a variety of glues and gums from their cell walls. These result in a film of micro-organisms bound into mucilage on the surface of the rock. Some mobile species can even move through this matrix. This *epilithic* film creates a remarkable boundary layer a few millimetres thick, providing a non-eroding laminar flow nearest to the community surface. This prevents the organic layer from being scoured from the surface by the molar action of the water.

2 Physical features such as the possession of heavy cases made of sand grains by caddis larvae may prevent dislodgement. The possession of a flattened body profile also helps organisms living under stones and in crevices, etc. The mayfly nymph *Rithrogena* has a very flattened body that allows it to live within the boundary layer on the tops of stones. Another factor is the streamlining of body form where the maximum body width is about one-third of the way along the body. This is common and best illustrated by fish such as trout and salmon.

3 Behavioural mechanisms are very important. Most species are positively rheotactic, e.g. the trout faces into the current, resulting in any locomotion tending to be upstream to counteract any downstream displacement. To facilitate living in protected regions under stones and in crevices, many species are negatively phototactic (making them move to dark places) and positively geotactic (making them move downwards to where such habitats might be). Another key behavioural pattern is positive thigmotaxis, which makes an organism maximise the amount of body surface in contact with a surface – try getting the highly thigmotactic earwig to cross a petri dish!

Despite these adaptations, displacement and drift downstream is common and in some cases is distinctly seasonal. For some insect larvae at least, it is part of the life cycle and is counteracted by the adult insects tending to fly upstream before laying their eggs.

4.2.1 Solar radiation

Radiant energy differs from other resources since, if it is not captured and used immediately, it is irretrievably lost. This contrasts with all other resources, which are recyclable. Once captured, it passes through the system only once. It comes as a flux of radiation from the sun after being modified by diffusion, reflection and absorption during its passage through the atmosphere. The amount arriving at the surface varies spatially and temporally depending upon the latitude (Figure 4.8). Direct radiation is greatest at low latitudes while seasonal variation is greatest at high latitudes.

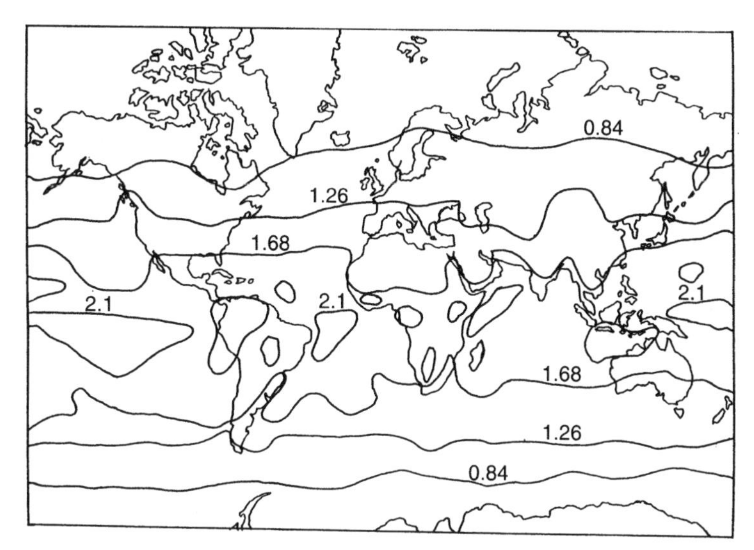

Figure 4.8 ***World distribution of solar radiation absorbed. Units are joules per sq. cm per min (adapted from Raushke et al. 1973)***

It is captured near the surface by plant leaves and, although much is lost by reflection and transmission, some of the light is absorbed and reaches the chloroplasts for use in photosynthesis. Solar radiation is a *resource continuum* in that it comprises a spectrum of different wavelengths (Figure 4.9a). However, only restricted wavelengths are usable for photosynthesis. All plants use chlorophyll to fix carbon and this molecule can only utilise wavelengths between about 400 and 700 nm. There are some other pigments such as bacteriochlorophyll which can absorb at different wavelengths but they are relatively unusual. The 400–700 nm band is called photosynthetically active radiation (PAR) and falls mainly within our visible spectrum. However, some 56 per cent of radiation reaching the Earth's surface lies outside this range and is unavailable. Aquatic plants receive their light filtered by the water column as well as the atmosphere. Water absorbs differentially (Figure 4.9b), removing the red, photosynthetically active wavelengths most rapidly and confining primary productivity to the upper layers of aquatic systems (the *euphotic* zone).

Variation in solar radiation may be systematic or random. *Systematic* variation derives largely from diurnal and seasonal rhythms. Such changes are regular and predictable and are used in many components of control processes in organisms that exhibit a determinate pattern of response. *Random variations* are associated with unpredictable events such as variations in cloud cover, shading, and reflection by waves. Because of this variation, no particular leaf morphology, physiology or orientation will be optimal all the time and strategic differences between plant species relate mainly to the evolution of 'sun species' and 'shade species'. Finally, it must be emphasised that

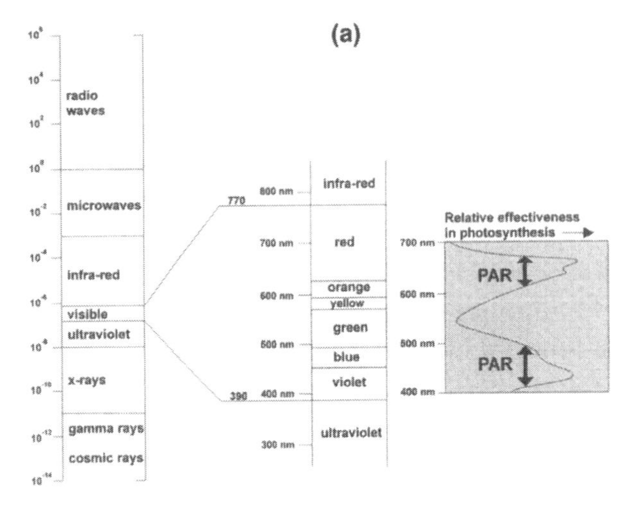

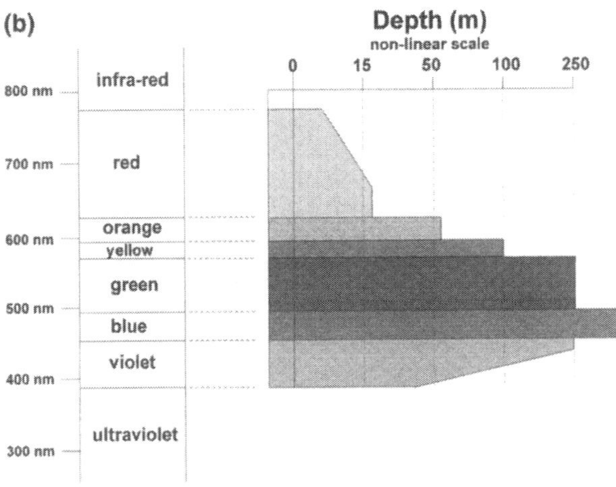

Figure 4.9 ***The biological significance of light. (a) The electromagnetic spectrum and the nature of the photosynthetically active radiation (PAR). (b) The relative penetration of water by different wavelengths of light***

there is a critical but complex relationship between the availability of water and the value of radiation as a resource. This is a consequence of the loss of water encountered when plant stomata (pores) open to permit carbon dioxide to become available for photosynthesis. There is, however, a potential conflict between the facilitation of photosynthesis and the conservation of water which becomes crucial when water is in limited supply. This has led to diverse adaptations to control water loss by plants (Box 4.5).

4.2.2 Inorganic materials

There are three inorganic resources that are vital because of their involvement in photosynthesis and primary production. They interact in a complex manner since radiation is used to split water molecules, reduce carbon dioxide and release oxygen.

4.2.2.1 Water

Water has already been considered as a condition (moisture, Section 4.1.2) which significantly affects the flow of water through a plant as transpiration. The water required for transpiration is effectively consumed and therefore, in this context, it is a resource. It is a vital resource for terrestrial animals and those living in hypertonic environments which, by having a higher osmotic pressure than the body fluids, may suffer from physiological desiccation.

All organisms require water as a medium for metabolism and, because of water losses through metabolic processes and leakage, there is a need for constant replenishment. Plants usually obtain water from the soil using their often elaborate and specialised root systems. Animals usually drink free water or obtain it from their food; some, such as kangaroo rats, use metabolic water only. Many habitats have limited water supplies that place significant restrictions on the distribution and abundance of larger animals, e.g. desert oases and waterholes.

Box 4.5

C3, C4 and CAM plants: different metabolisms for different conditions

Photosynthesis is such a fundamental process to life that it might be expected that it would be founded upon a single biochemical pathway. However, there is an intimate relationship between the availability of water and the 'type' of photosynthesis used. The basic problem lies in the fact that photosynthesis requires carbon dioxide, which is usually obtained by diffusion through the open stomata. However, open stomata lose water by evaporation that is fastest in warm conditions. This poses plants living in hot, sometimes dry, environments the problem of either restricting photosynthesis or suffering desiccation.

Three modes of photosynthesis occur.

- **C3** plants such as wheat, clover and oaks use a metabolic pathway called the Calvin (or Calvin–Benson) cycle to fix carbon dioxide directly. The first detectable product is phosphoglycerate (PGA), a three-carbon molecule, hence the name C3. However, photorespiration causes the loss of up to a half of their photosynthetically fixed carbon during the Calvin cycle in a temperature-sensitive reaction; in tropical habitats, where the temperature is often above 28°C, the problem becomes severe and limiting.
- **C4** plants such as corn (*Zea*), sugar cane and millet fix carbon dioxide by forming phosphoenolpyruvate (PEP) and then oxaloacetate, a four-carbon molecule, prior to using the Calvin cycle.
- Plants such as *Opuntia* spp. and other desert succulents with **CAM** (crassulacean acid metabolism) fix carbon by the four-carbon molecule route. However, they do so only at night when stomata can open with minimal loss of water.

Plants with 'normal' (C3) photosynthesis are relatively wasteful with their water compared with plants using the C4 or CAM methods. C4 and CAM plants can achieve significantly higher photosynthetic rates with less water loss and flourish where it is hot and dry. The water-use efficiency (carbon fixed per unit of water transpired) of C4 plants may be double that of C3 plants. The table below summarises the key differences.

Table B4.1 ***A comparison of the key features of the three plant metabolic types***

	C3	*C4*	*CAM*
Enzyme used	RubisCO	PEPCase	PEPCase
Optimum temperature	15–25°C	30–40°C	35°C
Leaf structure	bundle sheath cells lacking chloroplasts	bundle sheath cells having chloroplasts	mesophyll cells with large vacuoles
Efficiency in light	can be shade or sun plants	ineffective in shade	
Typical habitat characteristics	requires relatively moist habitats	arid or tropical regions	arid environments or where carbon dioxide is limited
Productivity (tons/hectare/yr)	22 +/– 0.3	39 +/– 17	low and variable

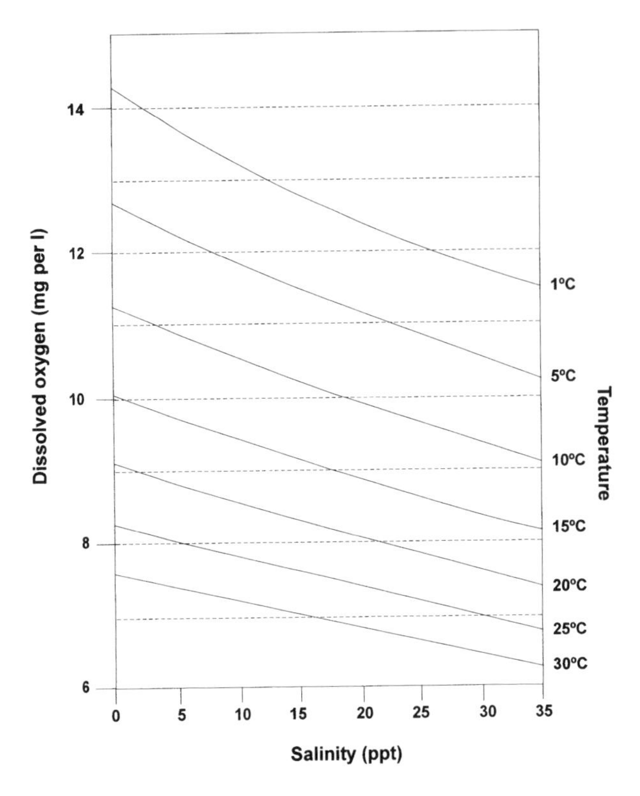

Figure 4.10 ***The effect of temperature and salinity on the amount of oxygen dissolved in water. Increasing salinity and/or temperature result in a significant decrease in oxygen availability***

4.2.2.2 Carbon dioxide and oxygen

Oxygen is a resource for both animals and plants and only a few prokaryotes are able to survive without it. It can become a limiting factor in many aquatic or waterlogged sediments because its solubility is relatively low and diffusion is slow. Its solubility in water is very temperature- and salinity-dependent (Figure 4.10). Because of its vital role in aerobic respiration, organisms in static environments may cause rapid local depletion, requiring ventilation mechanisms to provide a fresh supply of oxygen. Terrestrial organisms rarely suffer from a lack of available oxygen because of its relative abundance in air. However, in order to obtain atmospheric oxygen, they risk water loss because the respiratory surface can only function when moist.

Carbon dioxide is a vital resource for photosynthesis and is obtained readily from the atmosphere despite its concentration being only about 300 ppm (or 0.033 per cent of the air) but increasing (see Figure 3.14). It is fixed into living material either directly by terrestrial autotrophs or indirectly by aquatic photosynthetic organisms after it has dissolved into the water (Figure 4.11), where it exists in a variety of chemical forms. It diffuses freely in air and is limiting only in very rare circumstances. It is not, therefore, a source of competition between organisms.

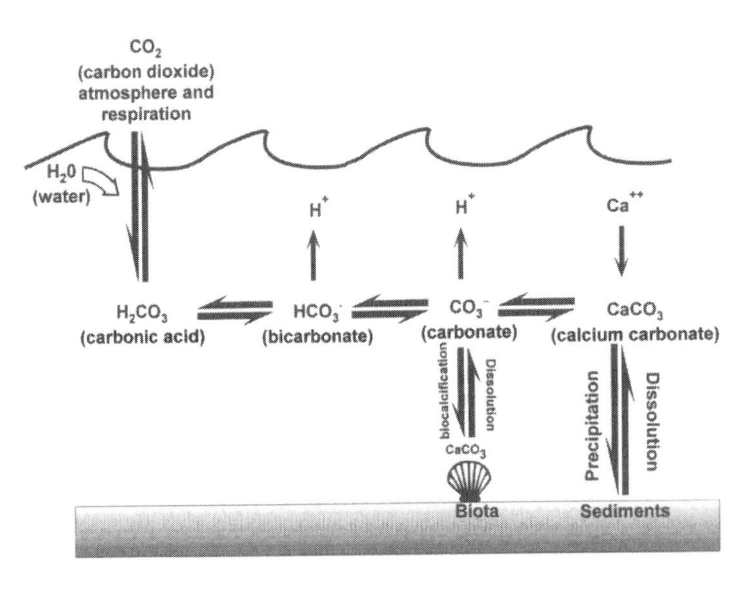

Figure 4.11 ***The dissolution of carbon dioxide in the water column and its subsequent interaction with sediments and the biota***

4.2.2.3 Minerals (as nutrients)

All organisms require a complex array of basic minerals to function properly. Animals obtain most of these materials from their food but plants must obtain them from the soil or the surrounding water. These include the *macronutrients* (required in relatively large quantities) such as nitrogen, phosphorus and sulphur (primary nutrients) and potassium, calcium, magnesium and iron (secondary nutrients). It also includes *trace elements* that are required in small quantities only. If any one of the macronutrients is in short supply, it becomes limiting but many of the trace elements are less critical in the short term. Because all plants require them, macronutrients are a major source of competition. Specialists need other minerals, e.g. silica is required for diatom skeletons. The extraction by plants of nutrients from the soil is a function of the root system and takes place alongside the water-gathering activities of the roots. Because minerals must be in a soluble form for absorption by the roots, there is a close relationship between water and minerals as resources: lack of water can make the minerals unavailable. Soils and water bodies differ markedly in their mineral status and this causes differences in plant distribution and abundance.

The stratification of aquatic systems can have very significant effects on nutrient recycling and, therefore, on the primary productivity of water bodies (Box 4.6).

4.2.3 Food

Food is a resource for all heterotrophic organisms and each organism is alternately a consumer and a prey item within food chains and webs. Three broad categories of feeding exist:

- *Decomposers* break down the dead remains and excreta of other organisms and recycle essential materials in the process.
- *Parasites* normally feed from only one or a few hosts during their lifetime and usually they do not kill their host.
- *Predators* usually kill and eat at least a part of the prey organism. Usually they feed from a relatively large number of prey items during their lifetime. Grazers are really a sub-category of predators. They often eat whole organisms (e.g. a limpet grazing a rocky surface) but may also eat only part of the prey without killing it (e.g. cattle grazing grass where the roots persist and regenerate new leaves).

These categories were discussed in more detail earlier (Section 3.2.1). The key point is that the more specialised the food resource/s required by an organism, the more the organism is forced either to live in patches of that resource or to spend time and energy searching amongst mixtures of organisms. Consumers may be *generalists*, with or without preferences within their food range and selecting from a wide variety of prey species, perhaps dependent upon availability, which may itself vary with time of year. *Specialists* feeding upon a single species or a restricted range of closely related species are often highly adapted in terms of morphology (particularly mouth-parts) and behaviour for the seeking and capture of that prey (Box 4.7).

Box 4.6

Thermoclines and productivity

Most lakes and oceans can be considered to have three distinct vertical layers (**stratification**):

- an upper, well-lit, well-mixed and relatively warm layer (**epilimnion** in lakes),
- a relatively thin intermediate layer which shows a rapid decline in temperature with depth (the **thermocline**) and
- a deep layer (**hypolimnion** in lakes) which is typically cold and dark.

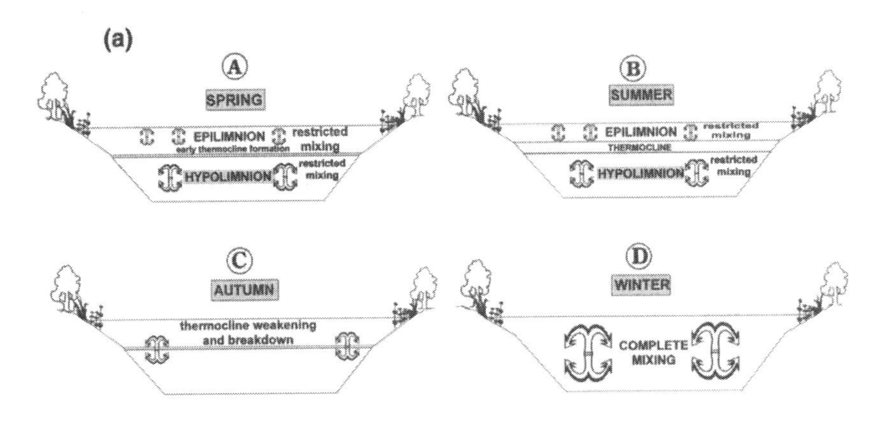

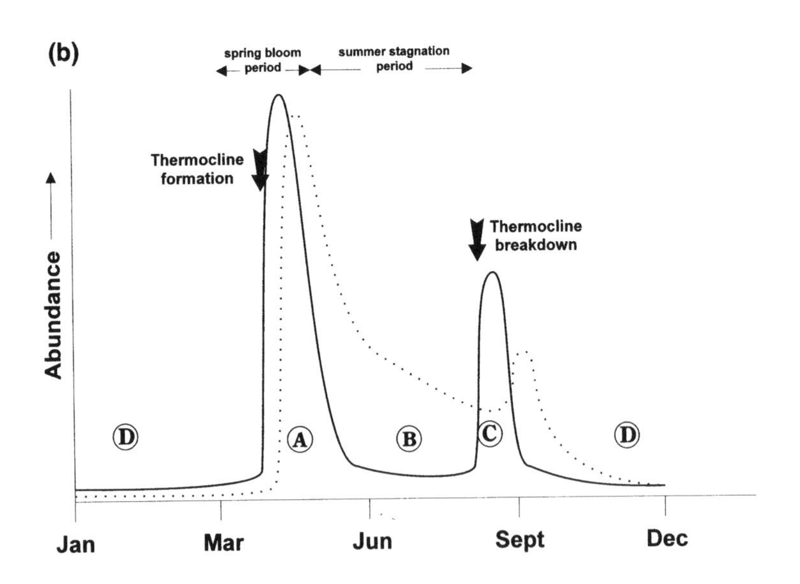

Figure B4.1 ***The seasonal cycles in a temperate region lake and their biological significance. (a) The changes in stratification of the water column. (b) The changes in the relative abundance of the phytoplankton (——) and zooplankton (.........) components associated with the changes in (a)***

Box 4.6 continued

The key structure here is the thermocline, which, although seasonal in character in many water bodies, effectively separates the upper and lower layers in terms of nutrient circulation. It thus has a dramatic effect on productivity within the water body.

The ocean thermocline is permanent at most latitudes and occurs at between 100 and 1000 m depth. In temperate lakes, the thermocline is highly seasonal and occurs at depths down to about 20 to 30 metres. It may be absent in shallow systems. Permanently stratified lakes are **meromictic** while those which stratify and destratify several times during a year are **polymictic**. The thermocline is a density layer, formed as a consequence of the thermal properties of water and water bodies, but, once formed, it can be remarkably stable. Major oceanic events such as El Niño are related to the effects of climatic factors on the permanent thermocline in the Pacific Ocean.

The significance of thermoclines is well illustrated by the seasonal cycles shown in temperate lake systems (Figure B4.1). The effect of the thermocline is to restrict the redistribution of nutrients to the warm, uppermost layers during the potentially productive summer months. This causes a summer stagnation period (of productivity) which is the result of the upper layers using up their nutrients and being unable to replenish them because of the restricted circulation.

The impact of thermoclines in low-latitude oceans can be overcome by oceanic phenomena such as **upwellings**. Here nutrient-rich deep water is drawn or pushed up into the potentially productive (warm and well-lit) but otherwise nutrient-depleted upper waters. This can lead to enormous, localised productivity increases such as occurred in Peru where a massive anchovy fishery was dependent upon such an upwelling.

Many food sources are seasonal and the nutritional value of a food resource may vary with season or stage of life. Some foods are rich in easily available energy while others, particularly plant materials with their cellulose cell walls, require major modifications and adaptations of the digestive system in order to extract nutritional materials. Because most animals lack cellulases (the enzymes required to digest cellulose), a necessary precursor to digestion of plant materials is chewing or grinding by a **gizzard**. This is a muscular region preceding the stomach which may contain small stones to assist the physical breakdown of the cellulose cell walls. Many herbivores have highly modified alimentary systems based around the strategy of utilising mutualistic, cellulolytic bacteria. The rumen and/or caecum of ruminant mammals such as cows and sheep are in effect temperature-regulated culture chambers for the microbial degradation of cellulose, the host absorbing the digested products. The bacteria receive a regular supply of food material and a safe environment for their own survival (Box 4.8). Carnivores do not have this problem of digestion of food materials but instead have difficulties of finding, capturing and handling prey items. This includes the development of foraging strategies to maximise the potential energetic returns of selective feeders feeding in mixed populations (communities)

A corollary of the utilisation of food resources is that those resources have often evolved physical, chemical or behavioural defences against potential consumers. This process often leads to coevolution and to increasing specialisation in the predators since they have to evolve to overcome the difficulties (Box 4.9).

Box 4.7

Bird beaks: the right tools for the job

Figure B4.2 ***Some examples of the diversity of bird beaks adapted for feeding on different materials***

Interspecific competition for available food resources is a powerful selective factor, illustrated very well in the consideration of bird beaks. Bird beaks have developed to allow specialised feeding and therefore resource partitioning of available food. It was Darwin's observations on the speciation of finches in the Galapagos Islands that first demonstrated this. The descendants of the few finches that originally colonised these geographically isolated islands were confronted by a variety of ecological niches which were unoccupied. As they spread into these niches, selective pressures led to the evolution of the thirteen species of Darwin's finches that can be grouped into three distinct types:

Box 4.7 continued

- *Ground finches*, each of which feeds on seeds of various sizes and has a bill whose size is related to the size of the seeds on which they feed
- *Tree finches*, whose bills are adapted for feeding either on buds and fruits or on insects of a specific size or from a specific habitat
- A *warbler finch* that occupies the niche normally occupied by mainland warblers, searching for insect prey on leaves and branches and trees.

An examination of the current types of bills possessed by the diversity of birds (Figure B4.2) illustrates very clearly how they have become adapted to different foods and feeding mechanisms. The relationship to resource partitioning is seen by examination of the bills of common European shorebirds (Figure B4.3), all of which feed in similar sandy or muddy areas and appear to occupy the same niche. However, their bills are of different lengths, thus allowing the exploitation of different components of the fauna within the sediments.

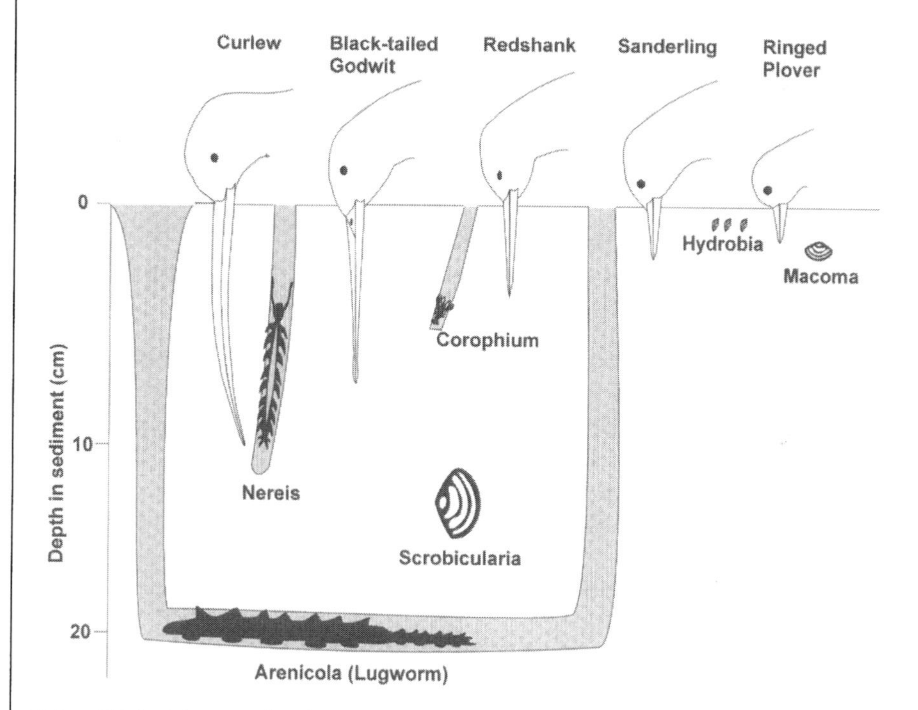

Figure B4.3 ***Diagram showing how the differing lengths of the beaks of estuarine wading birds facilitate the utilisation of different food species (resource partitioning) (modified from Green 1968)***

4.2.4 Space

All organisms require a space within which they can live. For a plant this may be only a small area but some top carnivores, either as individuals or groups (e.g. a pride of lions), may have a space requirement (territory) of many square km. Space is frequently a resource required because it contains other resources such as food, but many sedentary species require space simply because there is a real limit to the physical packing of an organism. Thus, barnacles that settle too densely will grow too tall to survive wave action since they are prevented from growing in breadth by the proximity of their fellows. Space may also be required for specific tasks such as breeding and hibernation and the availability of suitable nesting sites may be limiting for some bird populations, e.g. those requiring holes in dead trees.

Box 4.8

Ruminants and symbioses

Ruminant animals such as cows, sheep, antelopes and giraffes consume grasses, leaves and twigs rich in cellulose that mammals can't normally digest because they don't produce cellulases. Their survival is dependent on populations of symbiotic microbial populations contained in a specialised chamber called the rumen. This is in effect a culture vessel for growing populations of cellulose-digesting bacteria and protozoa. The rumen provides a relatively stable and uniform environment which is anaerobic, acidic (pH 5.5–7) and warm (30–40°C), optimal conditions for the microbial community. In the rumen, cellulose, starch and other materials are digested to form carbon dioxide, hydrogen gas, methane and low molecular weight acids, the latter being absorbed into the bloodstream for metabolic use. They also digest bacterial proteins but they expel the methane that then contributes to the greenhouse effect.

Rumination, the rechewing of previously ingested food, physically grinds food material to provide an increased surface area for microbial digestion. Ruminants are excellent users of low-grade, high-cellulose foods. The rumen harbours a great diversity of micro-organisms that include cellulose digesters, starch digesters, hemicellulose digesters, sugar fermenters, fatty acid utilisers, methanogenic bacteria, proteolytic bacteria and lipolytic bacteria, together with large populations of protozoans, mainly ciliates.

The microbial community of the rumen rapidly adapts to changes in the food type by changes in the relative proportions of the different component populations, a feature made possible by the highly diverse nature of this community. Non-ruminant herbivores such as horses and rabbits utilise similar microbial digestion within an enlarged structure called the caecum.

Box 4.9

Defences against predation

Plants and animals subject to herbivory and predation have developed defensive mechanisms to minimise the damage due to these processes. These act as a selective pressure on the consumer leading to the development of better or modified methods of consuming those items. This reciprocal evolutionary relationship is termed coevolution.

The defences fall into four major categories that are not mutually exclusive and combinations are common:

- *Physical defences* comprise structures such as spines, as exhibited by the leaves of holly trees and the spines of the stickleback and the hedgehog: they are a very common form of protection. Another common form is the development of protective coatings such as the outer coatings of nuts and other seeds, the bark of trees and the shells of molluscs, tortoises and turtles.

Box 4.9 continued

- *Chemical defences* are found commonly and many are used as pharmaceuticals. Plants contain diverse chemicals which appear to have no known role in plant biochemistry. They range from simple molecules such as oxalic acid (wood anemone) and cyanides (clover) to the complex glycosides, alkaloids, terpenoids, saponins, flavenoids and tannins. Such substances significantly affect grazing rates. In animals, secretions such as sulphuric acid (some gastropod snails), formic acid (ants) and quinone (Bombardier beetle, *Brachinus* sp.) are used to repel predators directly. Other animals such as poison frogs and many caterpillars accumulate toxic substances, often derived from their food, within their bodies, so that predators learn to avoid them.
- *Morphological defences* comprise mainly methods of hiding (**crypsis**) or of adopting warning coloration. Crypsis is shown by animals such as stick insects, the green coloration of many caterpillars and transparency of aquatic species. Such animals may be palatable but their cryptic morphology reduces their risk of predation. *Warning coloration* is the adoption of bright colours and patterns by noxious or dangerous animals, effectively advertising their dangerous nature. Bees and wasps have banded yellow and black markings to warn of their capability for stinging. This is known as **aposematism** and its effectiveness is demonstrated by another phenomenon, *Batesian mimicry*. Here, non-toxic animals adopt the coloration patterns of species which are dangerous and thus use the warning coloration to disguise their palatability. Thus harmless hoverflies mimic the coloration of bees and wasps.
- *Behavioural mechanisms* largely comprise avoidance or escape activities. Avoidance behaviours lead to organisms avoiding periods of predator activity: the daily vertical migrations of nekton and plankton may be for this purpose, as is the nocturnal behaviour of many species. Living in holes or burrows is also an avoidance strategy. Escape activities include fleeing but also include passive reactions such as 'playing dead', shown by a variety of mammals such as opossums and insects such as some beetles and grasshoppers. Some use sophisticated distraction methods such as the release of ink by squids. Others try to threaten the predator by appearing to make themselves larger, for example by possessing large eyespots, like many butterflies. Some actually do make themselves larger, for example the puffer fish, which enlarges itself by filling with water when threatened.

Summary points

- Conditions are a range of environmental factors that influence the distribution and abundance of organisms but which are never consumed or used up in the process. The adaptations and tolerances of the organisms determine their response.
- The main conditions are temperature, moisture, pH, salinity, current flow, soil, substrate/sediment structure and pollutants.
- Resources are environmental factors that are consumed or used up by organisms. They include solar radiation, inorganic molecules, organic materials (as food) and space. The desire for optimum exploitation of a resource by an organism leads to the development of foraging strategies.
- Each species occupies an ecological niche, a multidimensional (conditions and resources) concept of the totality of a species' requirements. No two species can occupy exactly the same niche but niches can overlap, giving rise to interspecific competition.
- Evolutionary adaptations occur in response to both conditions and resources and an adaptation by one species usually leads to a new responsive adaptation in associated species (coevolution).

Discussion / Further study

1 Pick an organism of your choice and find out as much as you can about its known requirements for conditions and resources. You might be surprised at how little we really know for most organisms.

2 How and why are organisms zoned with altitude? Construct a diagram to show the main biozones that occur from low-water mark on the shore to the snow-line in the mountains.

3 Select a major habitat of your choice and make a list of the microhabitats that you can identify within it. Consider how the conditions within the microhabitat are altered.

4 How would you expect temperature and oxygen to vary with depth in (a) a broad but moderately deep temperate lake in summer, (b) the same lake in winter and (c) a deep tropical lake?

5 Consider the relative advantages and disadvantages of being a generalist or a specialist in terms of a resource requirement such as food.

Further reading

Ecology, 3rd edition. M. Begon, J.L. Harper and C.R. Townsend. 1996. Blackwell Science, Oxford.
Chapters 2 and 3 are an excellent overview of the subject, well written and with plenty of examples.

Understanding our Environment: An introduction to environmental chemistry and pollution. R.M. Harrison (ed.). 1992. Royal Society of Chemistry, Cambridge.
A useful source for more detailed information about conditions.

Environmental Science: Working with the Earth, 5th edition. G.T. Miller. 1995. Wadsworth Publishing Co., Belmont, CA.
Useful and well-written overview of resources, particularly those associated with human society and our exploitation of the environment.

Environmental Data Report. United Nations Environmental Programme. 1987. Basil Blackwell Ltd, Oxford.
Compiled data sets for many conditions and resources, orientated towards human interests.

'Our fathers' toxic sins'. W. Stigliani and W. Salomons. 11 December 1993. *New Scientist* 140, 38–42.
Very interesting article on the storage and release of toxic chemicals in soils and the concept of chemical time bombs.

Beyond Silent Spring: Integrated pest management and chemical safety. H.F. van Emden and D.B. Peakall. 1996. Chapman and Hall, London.
An in-depth consideration of the problems and possible solutions of chemical use and integrated pest management.

References

Chapman, J.L. and Reiss, M.J. 1992. *Ecology: Principles and applications*. Cambridge University Press, Cambridge.

Elton, C. 1927. *Animal Ecology*. Sidgewick and Jackson, London.

Green, J. 1968. *The Biology of Estuarine Animals*. Sidgewick and Jackson, London.

Hutchinson, G.E. 1957. 'Concluding remarks', *Cold Spring Harbor Symposium on Quantitative Biology* 22, 415–27.

Lewis, J.R. 1964. *The Ecology of Rocky Shores*. English Universities Press Ltd, London.

Liebig, J. 1840. *Chemistry in Its Application to Agriculture and Physiology*. Taylor and Walton, London.

Raushke, E., Haar, T.H. von de Bardeer, W.R. and Pasternak, M. 1973. 'The annual radiation of the earth–atmosphere system during 1969–70 from nimbus measurements', *Journal of Atmospheric Science* 30, 41–346.

Shelford, V.E. 1913. *Animal Communities in Temperate America*. University of Chicago Press, Chicago.

5 Individuals

Key concepts

- **Organisms can be divided into unitary and modular forms, with the concept of an individual difficult to apply to modular species.**
- **The key themes in ecological physiology are resistance, tolerance and adaptation to environmental conditions.**
- **The key energy strategies are ectothermy and endothermy.**
- **Sedentary and mobile life strategies have considerable ecological consequences.**
- **Behavioural mechanisms are important in responding to environmental conditions and to resource acquisition.**
- **Asexual and sexual reproduction strategies determine the patterns of variation and population increase.**
- **Life cycles and life histories represent evolutionary responses to environmental factors.**
- **Feeding strategies and mechanisms are fundamental components of ecological relationships.**

Ecological studies mainly deal with the complexities of interactions within or between populations, communities or higher groupings. However, the fundamental units are the individuals whose biology is shaped by natural selection acting on species populations. It is the responses of these individuals to environmental conditions and resource requirements that form the basis of the more complex interactions at population and community levels discussed in subsequent chapters. This chapter reviews some of the main factors that operate at individual level and the ways in which they may influence the **autecology** of a species.

5.1 What is an individual?

Most ecological studies have been founded on the concept of individuals who are born, grow, die, emigrate and are immigrants, i.e. assuming that an individual is an easily identified species unit. Each species has a variable **genotype** (genetic content) despite the **phenotypes** (the outward expression of the genotype after interaction with the environment, i.e. appearance and physiology) being similar in most respects (see Section 6.1 also). However, it is important to remember that:

- all species pass through a number of different life cycle stages, e.g. egg, embryo, maggot, pupa, fly. Each stage is subject to different factors that influence the key processes of growth, reproduction and survival and each stage needs separate consideration.

- within any particular stage, individuals vary significantly in their *condition* with particular respect to factors such as nutritive status, parasitisation and size.
- the concept of the individual is difficult to apply to organisms which reproduce asexually to produce often colonial, clonal (genetically identical) forms. It is important to distinguish between these *modular* organisms and the more familiar and well-studied *unitary* individuals.

What are the differences between unitary and modular organisms?

5.1.1 Unitary and modular organisms

Unitary organisms have a highly determinate form that is not modified by environmental conditions (Table 5.1). Thus all spiders have eight legs, insects six and mammals four: similarly, mammals have only two eyes and birds have two wings. Their pattern of development and final form are predictable as is their life cycle, in sequence if not in timescale. The human species is a good example since our life cycle is clearly sequential, going from fertilisation to zygote and then to a foetus which exhibits all the basic unitary human characteristics. The sequence continues through infant, child (juvenile), adolescent, sexually mature adult and senescent old age stages, followed finally by death. Thus both form and sequence are determinate. The focus of much ecological attention has, for a long time, been dominated by unitary species. The majority of population studies concentrate on the dynamics of unitary organisms where the recognition of an individual is relatively straightforward.

Table 5.1 ***Overview of the main modular and unitary groups of organisms***

Modular organisms	*Unitary organisms*
many colonial protists such as algae and moulds	many solitary protists such as ciliates and many diatoms
most (colonial) fungi	fungi such as toadstools
most multicellular plants including grasses and trees	
colonial cnidarians such as corals sponges	solitary cnidarians such as most jellyfish and sea anemones ctenophores (sea gooseberries)
bryozoans (sea mats) and other colonial 'minor' phyla	rotifers, arrow worms, pogonophores and other solitary 'minor' phyla
	worm phyla such as Platyhelminthes (flatworms) and segmented worms (Annelids)
	arthropods (insects, crustaceans, etc.)
	molluscs
	echinoderms
hemichordates such as colonial tunicates (sea squirts)	hemichordates such as solitary tunicates (sea squirts)
	chordates including the vertebrates

Modular organisms are quite different. They are constructed from units (modules) which are themselves often capable of producing similar modules. The organism is usually *sessile* (non-mobile) and branched in the adult form, although there is a wide range in the numbers and types of each basic module, which is characteristically sensitive to environmental variables. The most familiar modular organisms are the higher plants, whose leaves are photosynthetic modules and flowers are reproductive modules. However, there are some nineteen modular animal phyla (Table 5.1) and modular organisms form the major components of many aquatic and terrestrial ecosystems. The key ecological feature of modular organisms is that the usual parameters associated with populations of unitary individuals (see Chapter 6, populations) cannot easily be applied to modular forms. For such forms, the equivalent of the unitary individual is the **genet** or genetic individual, the product of the zygote, which is made up of the repeated genetically identical units or *modules*. Thus a tree is a genet made up of leaf modules, flower modules, etc. and one can study either the distribution and abundance of genets, or of the modules, or of both. It is the abundance and distribution of modules rather than the genets that is most significant ecologically. The potential for variation (plasticity) in modular organisms is much greater than in unitary ones. The critical processes of birth, development, reproduction, senescence and death occur not only at the level of the genet but also at module level. The age structure of a population can be studied, therefore, at two levels: that of the genets and that of the modules. Thus a deciduous tree sheds its leaf modules each autumn, producing an important annual input of organic litter to the environment, while the genet survives to produce new modules in the following spring. In this way, the genet effectively avoids the programmed senescence typical of unitary species and death usually results from external factors such as disease or instability instead: the giant redwoods (*Sequoia* sp.) are a good example, some living for more than 2000 years.

5.2 Ecological physiology: resistance, tolerance and adaptation

Each organism is in a state of dynamic equilibrium with the environment, which is itself in a state of almost constant flux. All organisms must, therefore, be able to make internal adjustments in response to these external changes not only to survive, but to grow and reproduce the species. This is the realm of ecological physiology (ecophysiology). Organisms normally survive environmental variations by one or more of the following responses:

- by *resisting* or *tolerating* the changes when they occur, a process with distinct limitations for both the degree and the duration of the changes
- by establishing new equilibria through the *regulation* of their internal environment or by adapting their physiology to the conditions (*adaptation*)
- by *moving* (migrating) to a more suitable environment (see Section 5.4).

Organisms occupy an enormous diversity of habitats, from the blue-green algae (Cyanobacteria) of the hot thermal springs to the mammals and birds of the frozen polar regions and from the plants and animals of arid deserts to the unusual forms living in the cold, dark ocean abyss. This demonstrates the remarkable ability of organisms to evolve tolerance and compensation mechanisms to cope with diverse environmental conditions. The development of body organisations in which extracellular fluids bathe the cells to provide a barrier between the environment and the cells was an important evolutionary development. It allowed the evolution of mechanisms to control the composition of the fluids surrounding the cells, thus allowing the development of greater resistance and tolerance. This was particularly important during the evolutionary transition from the relative stability of the sea to freshwater and terrestrial environments.

Intracellular and extracellular fluids are often maintained at relatively constant ionic and osmotic compositions even when the external conditions fluctuate considerably. This process of maintaining internal constancy is **homeostasis**. The more complex unitary animals such as birds and mammals show this best, since in these groups the evolution of internal systems for control and regulation of the key physiological processes is most developed. The evolution of homeostatic mechanisms was essential to the development of diversity of form and function and to the colonisation of freshwater and terrestrial environments. It also allowed biochemical systems to become more efficient by adapting to operate within relatively narrow ranges of physico-chemical variables. When the external environment changes, there are two basic patterns of physiological response:

- **conformation**, where the internal variables fluctuate directly with variations in the environment and survival depends upon cellular resistance to the changes
- **regulation**, where internal variables are maintained at levels different from that of the environment, albeit at a significant energy cost.

The appropriate descriptive term is produced by adding a prefix to the variable name – hence for temperature the relevant terms are thermoconformation and thermoregulation. Organisms that are able to survive over wide ranges of an external variable such as temperature are termed *eury*thermal; the prefix eury- is applicable to other variables, e.g. euryhaline (salinity). The prefix *steno-* applies to organisms whose tolerance ranges of a variable are very small, e.g. stenothermal and stenohaline organisms. Another important terminology relating to constancy is the distinction between *poikilo*thermic forms, those with variable internal temperatures, and *homoio*thermic forms which have relatively stable internal temperatures. The prefixes poikilo- and homeo- apply to other physiological variables also (e.g. poikilo-osmotic). Animals that obtain their body heat from the environment are termed *ectotherms* (Figure 5.1) while those which have internal mechanisms for heat production (warm-blooded animals) are *endotherms* (see also Section 5.3).

5.2.1 Resistance and tolerance

All organisms exhibit responses to *stress*, the general term given to the physiological condition where the level of an environmental condition or resource is outside the optimum range for that organism. Figure 4.1a shows a generalised response curve to a

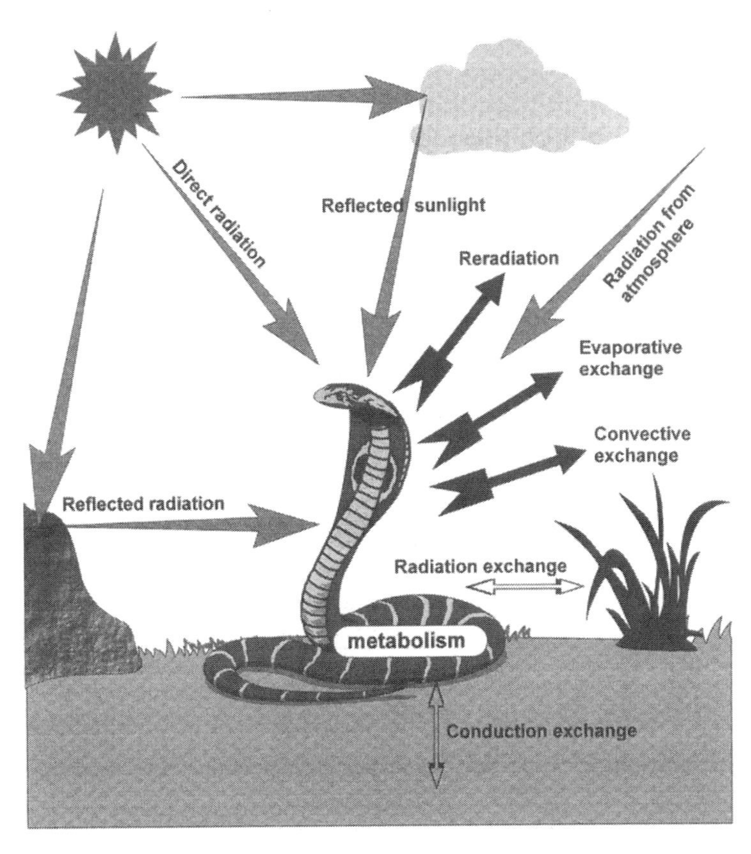

Figure 5.1 ***Schematic diagram of the factors which influence the heat balance of ectotherms***

generalised stressor. The shape of this curve for any given stressor will vary with species, life stage, season and physiological condition and it will also vary with the particular stressor. The range of any specific variable that an organism can survive indefinitely is its *Zone (Range) of Tolerance* (or Compatibility). Above and below this range are the *Zones of Resistance* in which the organism will eventually die as a result of that factor alone. The transition point between these two zones is the *incipient lethal level* (upper and lower) but this is very difficult to characterise since many factors cause it to vary. Response curves are also very difficult to define, however, even for a particular species since within a population there is considerable genetic diversity, resulting in physiological diversity and plasticity between individuals. Remember too that responses to environmental stressors, although often assessed experimentally in single-variable experiments, are frequently complex, with multiple interactions in natural situations.

Tolerance is largely a physiological property while *resistance*, although including physiological response, is significantly influenced by morphological and behavioural modifications. These modifications have evolved in response to potentially deleterious environmental conditions and are alternatives or supplements to the physiological homeostatic mechanisms. This is best illustrated by studying adaptations to extreme environments such as deserts, wave-splashed rocky shores and arctic/antarctic habitats. In deserts, cacti have reduced their leaves to spines to reduce the loss of water through stomata, while animals may burrow into cooler layers at the height of the day (Box 5.1). Animals in cold environments develop thick, thermal insulation (fur or fat), while on wave-splashed rocky shores organisms possess strong attachment devices and tough external materials (Box 4.3). Such evolutionary adaptations have allowed organisms to populate highly stressful environments where conditions are tough but interspecific competition is usually reduced.

Box 5.1

Life in deserts: water conservation and heat tolerance

Deserts are very dry areas found in both temperate and tropical regions, characterised by low densities of life and low diversity. The low humidity of the desert atmosphere results in very wide daily temperature fluctuations, with cold nights and extremely hot days. This is a highly stressful environment and its unique condition of dry heat requires highly specialised morphological, physiological and behavioural adaptations. Organisms must both conserve water and control their temperature, conflicting demands since many organisms normally use evaporative cooling for temperature control.

Plant cover is sparse but includes both perennial (cacti, Joshua trees, yucca and grasses) and annual species, all of which tend to have reduced leaves or no leaves, reducing transpiration and conserving water (*xerophytes*). Succulent plants such as cacti, which store water, grow continuously and often secrete a thicker layer of cuticle over the surface to reduce water loss. A dense covering of epidermal hairs is also common and reduces the drying effects of air currents. Annual plants often adopt a strategy of carrying out their whole life cycle during the brief period when water may be present, evading the dry periods as desiccation-tolerant seeds. To reduce competition for water, some plants secrete substances from their roots or shed leaves which inhibit the establishment of other plants nearby (*allelopathy*). They frequently possess spines, thorns and toxins to minimise grazing by herbivores. See Box 4.5 for a consideration of the ways in which photosynthetic pathways are modified for life under conditions of water shortage.

Desert animals tend to be small and most remain under cover or in burrows during the day, being active at night when it is cooler. Snakes and lizards are relatively common in deserts because their heavy skin prevents water loss and they produce mainly dry excretions. Most of their water derives from prey and from metabolic water, water produced during normal metabolism. Another famous example of an animal that can survive on metabolic water alone is the kangaroo rat. Larger animals living in arid regions such as camels and ostriches have heavy insulation (fur and feathers respectively) on their backs to keep out the heat from the sun. They are also physiologically very tolerant of dehydration and can lose up to a quarter of their body water and yet make it up in a few minutes when water becomes available. Many desert animals also have unusually large tolerances of body temperature change, thus reducing the need for evaporative cooling and conserving water.

5.2.2 Adaptation

The ability of a species to survive either competition or a particular set of complex environmental variables is the result of evolutionary changes over time (*adaptation*). Many responses of an organism in a fluctuating environment are adaptive since they enable the organism to survive and thus reproduce. The environment (physical, chemical and biological), therefore, acts as a selective force on a population of a species so that there is an evolution of mechanisms to facilitate survival. This is *evolutionary adaptation* and is genetic. Within the existing genome of any individual, however, morphological and physiological plasticity (phenotypic variation) permits adaptation to short-term environmental changes. This is non-genetic or *environmentally induced adaptation*.

A particularly important aspect of adaptation is the short-term adjustment of the physiology of an organism to the current conditions, a process known as *acclimation*. This is demonstrated when humans adapt to tropical temperatures over the period of a two-week holiday. Initially, they sweat profusely but by the end they don't; they have physiologically adapted to the temperature of the new habitat. This ability to adapt is a reflection of the physiological plasticity built into our genotype and it operates within the Zone of Tolerance. This adaptive ability is very important in experimentation since pre-experimental conditions can exert a significant influence over experimental performances and acclimation is a vital part of experimental technique.

5.3 Metabolic rate

The metabolic rate of an organism is a measure of the amount of energy used per unit of time and reflects the relationship between the individual and its environment. Organisms require energy for diverse processes such as growth, reproduction, body maintenance and locomotion. Some highly active forms such as birds have very high metabolic rates while others such as crocodiles are able to conserve energy and survive with relatively low metabolic rates. The **basal metabolic rate** of an organism is the minimum that will allow survival but is not a very valuable measure ecologically. Many factors affect the metabolic rate and a more meaningful measure is the *daily energy expenditure*. This includes the energy spent on stress regulation, growth, reproduction and locomotion and typically is about one and a half to three times the basal metabolic requirement.

Diverse factors modify the metabolic rate of an organism but these mainly comprise interactions between lifestyle factors and body size. The most significant lifestyle factor is the means by which the organism obtains the thermal energy necessary for its life processes. *Ectotherms* acquire heat energy from external thermal sources (sunlight usually) and are the most widespread type of organisms, ectothermy being considered to be the most ancient strategy. They are often referred to as *poikilothermic* ('cold-blooded') organisms, whose internal temperature is allowed to vary and which include animals such as fish. An important consequence of this strategy is that such organisms often have periods of lethargy associated with low body temperatures. *Endotherms*, however, use energy from their food to heat their bodies and regulate their internal temperatures precisely, providing advantages in cell biochemical activity as a result of the thermal stability. They are *homoiothermic* ('warm-blooded') organisms that maintain a constant internal temperature. Note, however, that some ectotherms such as lizards can control internal body temperature to within about 5°C by adopting particular behavioural strategies. Homoiotherms generally have basal metabolic rates some 25–30 times greater than similarly sized poikilotherms, reflecting the amount of energy needed to maintain body temperature.

Size, like body mass, has an important influence on metabolic rate. Figure 5.2a shows the basal rates of a variety of species as a function of their body mass. Generally, within any particular group of organisms, larger organisms need more

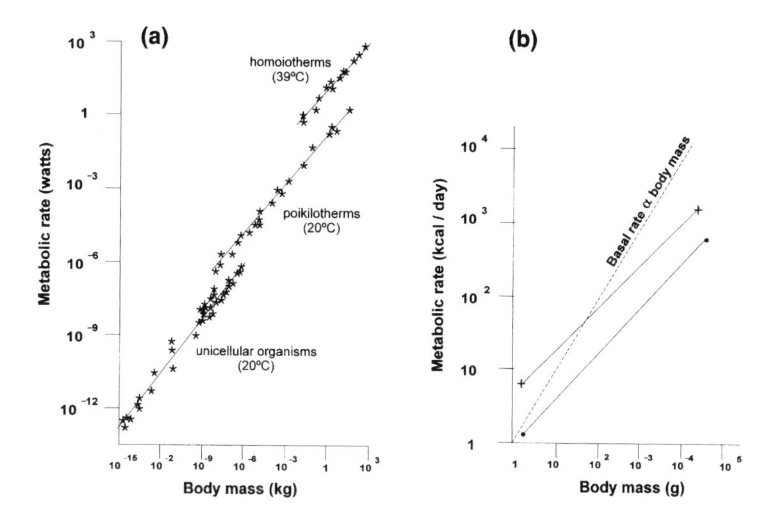

Figure 5.2 ***Metabolic rate and size. (a) Basal metabolic rates of three groups of organisms as functions of their size. (b) Daily energy expenditure (+ — +) and basal metabolic rate (o — o) based on a study of 47 bird species (adapted from Bennett and Harvey 1987)***

energy each day to live but the relationship is not directly proportional (Figure 5.2b). Larger organisms actually need less energy per day than smaller organisms, *relative to their body mass*. Therefore, the smaller the organism, the higher the metabolic rate tends to be.

Ectotherm strategy is based on a *low-energy system* (Pough 1983), where the objective is to heat the body into a temperature range suitable for foraging actively and as efficiently as possible. When conditions or resources are limiting, they can switch off as energy consumers for extended periods by going into a *torpor*: when they are favourable, ectotherms become active and grow and reproduce very efficiently. The energy saved by not maintaining body temperature is used for other body requirements. This strategy results in several characteristic features:

- a low resting metabolic rate
- a high production efficiency
- periods spent in a torpor
- maximum exploitation of episodic favourable conditions
- the possibility of small size without the endothermic metabolic costs
- the possibility of vermiform or serpentine shapes with large surface:volume ratios without the large energy losses that would occur in an endotherm.

The abundance and diversity of ectothermic forms reflect their domination of the biological world, yet endotherms are often thought of as being more successful. What are the advantages conferred on endotherms, therefore, that have made this alternative strategy effective for so many higher animals? Endothermy is a *high-energy system* (Pough 1983) which uses energy from cellular metabolism to maintain an elevated and constant body temperature. This provides endotherms with ecological advantages since they become significantly independent of environmental conditions. It allows the exploitation of habitats and conditions that are effectively denied to most ectotherms, e.g. foraging at night or in the shade in cold climates and inhabiting high latitudes. Nocturnal ectotherms are mainly confined to warm climates, unlike endotherms. The cost lies in a lower production efficiency since much of the energy acquired by feeding has to be diverted to the requirements of endothermy and is therefore not available for incorporation into body growth or reproduction.

Another major cost to be met from the energy budget of an organism is locomotion (Box 5.2). The energy cost for any given form of locomotion is directly related to the mass to be moved. It is interesting that swimming and flying have been found to use less energy than running.

Box 5.2

Locomotion and Reynolds numbers: life in different worlds

When an organism moves, it is subject to two fundamental forces, inertia and drag. Once in motion an organism has inertia, the property of matter that causes it to resist any change in its motion. However, while organisms in motion have inertia, they also experience drag, a force resulting from the viscous properties of the medium, which tends to resist the movement of objects. The relative significance of each force is directly related to the size of the organism, its velocity and the viscosity of the medium through which it is moving. This is summarised as the Reynolds number (Re), a dimensionless number expressing the ratio of inertial to viscous forces. Where Reynolds numbers are large, inertia is much more important than viscous forces while at small Reynolds numbers, inertia is unimportant and viscous forces dominate. A dragonfly moving through the air at 7 metres per sec has a Re of about 30,000 while for a planktonic ciliated larva in the sea it may be only 0.3. Generally, because water is much more viscous than air, aquatic organisms have lower Reynolds numbers for similar velocities and sizes. Why is this important?

The significance of Reynolds numbers lies in their consequences for locomotion, ciliary feeding and the sinking rates of small organisms in aquatic environments. We live in a world of high Reynolds numbers, dominated by inertia forces, and where drag, although noticeable while we are swimming or driving a car, is of relatively minor significance. However, pity the poor aquatic ciliate living in a world of low Reynolds numbers, where inertia is insignificant and viscous forces dominate – moving would be like trying to move through treacle. To move under these conditions requires doing a lot of work and expending a great deal of energy; when locomotory activity ceases, so does movement – instantly. There is no such thing as gliding at low Reynolds numbers. However, it is not all bad. A very positive feature is a reduction of sinking rates. Therefore, the planktonic mode of life benefits from this dominance of viscous forces, especially when combined with mechanisms for reducing the density of the organism to match that of the water. Increasing the surface area maximises drag and dramatically reduces passive sinking rates: this is why so many planktonic forms have spines or thread-like body extensions. Aerial plankton has to be very small because air is less viscous than water. Very small insects such as some midges have hairy wings and effectively row themselves through the air rather than fly: at their size the air is viscous enough to support them.

A final important point is that the movement of water by cilia is also a very energy-consuming process because the cilia are working at low Reynolds numbers and are effectively moving a very viscous substance. The energetic significance of this for organisms using ciliary-generated feeding or respiratory currents is obvious.

Many parameters need consideration when an energy budget for a particular species is being constructed. Individuals cannot devote all their resources to reproduction alone; they use some of their energy intake for maintenance, growth and living costs. When an organism ingests food, only a proportion is incorporated into that organism (assimilation). The term ‘assimilation efficiency’ describes the percentage of the ingested food that is assimilated rather than egested and this varies with both the organism and the food type. It is as high as 90 per cent in carnivores feeding on vertebrates and between 30 and 60 per cent in most herbivores, because much of the plant material is indigestible. However, it can also be much lower than 20 per cent, especially in invertebrate deposit feeders.

Once assimilated into the body, food materials are either respired to release the energy needed for metabolic activity or used for growth and reproduction. Growth and reproduction are collectively termed *production*, a very important ecological parameter since it represents the material available to the next trophic level. The percentage of assimilated energy which an organism diverts to growth is its *growth efficiency* while that diverted to reproduction is the *reproductive efficiency*. Growth efficiency is usually higher in the younger life stages and diminishes with increasing size and maturity: any factor affecting the utilisation of energy will influence this measure. Thus non-mobile, juvenile poikilotherms will tend to have higher growth efficiencies at between 50 and 80 per cent. Bacteria and protists in culture have growth efficiencies of about 60 per cent. Reproductive efficiencies are rarely studied but reproductive strategies clearly have a major influence. Thus, an annually reproducing species will clearly have to allocate more energy and materials to reproduction each year than a species which reproduces only once during its lifetime.

5.4 Behaviour

Individuals have to carry out diverse functions to survive and reproduce. Patterns of activity have evolved around these requirements to include obtaining food (see Section 5.7), avoiding predation and adverse environmental conditions, and finding a reproductive partner and associated resources where necessary. All species capable of movement can respond to environmental variables. Plants and fungi usually respond by growth mechanisms while many animals have the advantage of more rapid locomotory responses. Variations in these modes of behavioural response have important consequences for the organisms involved and, of course, have differing energetic consequences and requirements. This section introduces some broad aspects of behaviour as it affects the response of organisms to environmental conditions and the resulting patterns of distribution and abundance. However, any aspect of behaviour is of ecological significance for each species but limitations of space preclude a detailed discussion in this book.

5.4.1 Sedentary or mobile – the strategic consequences

All organisms, whether sedentary or mobile, exhibit behavioural responses to the environment. Sessile organisms have opted for a low-energy system with risks. Remember that plants and most fungi are sedentary but can still respond to the environment by differential growth. The definition of mobility, therefore, involves a consideration of timescale, as can be shown using time-lapse photography, but mobile organisms are usually considered only to be those capable of a locomotory response. The main consequences of a sedentary lifestyle are:

- Protective mechanisms must be developed to avoid predation because escape behaviour is not an option.
- Damaging environmental conditions must either be tolerated or a behavioural mechanism for selecting stable habitats must be developed. This must operate at the pre-settlement stage.
- Food capture is dependent upon the environment carrying food to the vicinity of the organism, e.g. by water currents. This is not a problem for photosynthetic organisms but can be for organisms at other trophic levels.
- Asexual reproduction is straightforward but sexual reproduction involving cross-fertilisation poses significant problems. *Broadcast fertilisation* enables genetic mixing but is costly in terms of energy and materials. *Copulation* requires a gregarious settlement behaviour to ensure that a partner is within the range determined by its copulatory system, e.g. barnacles.
- Dispersal must take place at an early stage of the life cycle and must avoid the competitive disadvantages of over-dense settlements, i.e. intraspecific (within species) competition.

A mobile lifestyle, however, is a high-energy system which avoids many of the difficulties associated with a sedentary lifestyle because it allows a flexible response. However, it does require an appropriate body design to provide locomotion. Its main advantages are:

- Predation can be avoided by the development of escape mechanisms enabled by locomotion.
- Damaging environmental conditions may be avoided by appropriate behaviour. This can be small-scale, using taxic behaviour (see Section 5.4.2), or large-scale involving migration over large distances (Box 5.3).
- Food can be searched for and a wider range of food becomes available.
- Sexual reproduction by copulation becomes more practicable, reducing the dependence upon wasteful broadcast fertilisation.
- Dispersal can occur at any life stage in which the organism is mobile and competitive factors reduced by mobility and redistribution.

Thus, the evolution of differing lifestyles has diverse and important potential ecological consequences.

5.4.2 Behavioural mechanisms

Individuals must respond to the environment, both biotic and abiotic. This behavioural response should modify the relationship between the organism and its environment to favour the survival of the organism. All living organisms exhibit a variety of behavioural activities determined by the extent to which they are able to respond to stimuli. Plants, like animals, need some form of co-ordination if their growth, development and survival are to be optimised in response to environmental changes.

Box 5.3

Avoidance and dispersal

There are many stress situations that can be avoided by the adoption of a suitable strategy operating in space or time. *Spatial* avoidance and dispersal is clearly most easily visible in animals which are mobile. Even in sessile species of animals and plants, however, the adoption of dispersal mechanisms which ensure that future generations are not subject to competition with parental organisms is a form of avoidance behaviour. *Temporal* avoidance and dispersal occur in variable environments where periods of adverse conditions or resource shortages are avoidable by using life-cycle strategies such as diapause, hibernation, aestivation or migration.

The onset of adverse conditions, the failure of a resource, or changes in intraspecific or interspecific competition all require an appropriate response. This may take place at individual or population levels although it always operates through the response of individuals. Such activities may be very localised, e.g. simply moving from one tree to another to exploit new resources or hiding from adverse conditions by sheltering in a burrow or other microhabitat. However, in those parts of the biosphere where changes are regular and either predictable (e.g. tidal phenomena) or unpredictable (time of onset of snows or rainfall), these avoidance activities can become programmed into the behavioural ecology of the organism. Good examples are:

- diapause periods in many insects involving the use of a resting phase in the life cycle which is intended to avoid unsuitable environmental conditions.
- hibernation periods, as in mammals such as bears and squirrels, where physiological changes occur to reduce metabolism during adverse periods; the animals effectively sleep through the poor conditions while safely hidden in some protected microhabitat.
- diurnal and tidal movements in which populations move between habitats repeatedly on timescales of hours, days or months. Examples are the daily vertical migrations of ocean plankton and nekton to exploit different food resources, and the daily movements of bats and owls, moving between sheltered roosts and feeding grounds.
- seasonal migrations such as the massive movements of wildebeest in Africa in response to seasonal changes in conditions and resources. Also, the remarkable migrations of birds such as the Arctic tern, *Sterna paradisea*, which performs a trans-global migration each year from one pole to the other and back, a round-trip of about 40,000 km: it breeds only in the north.

The terms 'dispersal' and 'migration' describe various aspects of the mass movement of organisms and it is important to remember that there is no sharp distinction between these processes.

Behaviour, therefore, varies from the relatively simple growth of a plant stem towards light to the complex behavioural patterns associated with sexual reproduction in birds and mammals (territoriality, courtship and mating). All behaviours can be broadly grouped into actions intended to avoid environmental threats, to find food, to reproduce, or to interact (communicate) with members of the same and other species.

- **Plants** do not possess nervous systems and rely entirely on chemical co-ordination. Their responses are therefore slower than those of animals and often involve differential growth rather than true movement. They show two main types of response:

1 **Tropisms**: growth movements in response to, and directed by, an external stimulus which can be light, gravity, water, chemicals or touch (Table 5.2). These are positive when the response is towards the stimulus (e.g. the growth of a plant towards light is positive phototropism) or negative when away from it (e.g. the upward growth of a shoot is negative geotropism).

2 **Nasties** (singular nasty): a non-directional movement of part of a plant in response to an external stimulus (e.g. light intensity, temperature or touch). The structure of the responding organ determines the direction of movement. Good examples are the closing of some flowers at night and the rapid collapse response of the leaves of the sensitive plant, *Mimosa pudica*, when touched. Insectivorous plants show some of the most elaborate nastic movements.

Table 5.2 ***Plant tropisms***

Type of tropism	*Stimulus*	*Examples*
Chemotropism	chemicals	Fungal hyphae may be positively chemotropic and pollen tubes respond positively to substances produced by the ovule micropyle.
Geotropism	gravity	Shoots are negatively geotropic while roots are positively geotropic.
Phototropism	light	Shoots are positively phototropic and some roots are negatively phototropic.
Hydrotropism	water	Roots and pollen tubes are positively hydrotropic.
Haptotropism	solid surface or touch	Tendrils of climbing plants and sensitive structures of insectivorous plants are positively haptotropic. The most famous example is the leaves of *Mimosa pudica*, the sensitive plant.

- **Animal** behaviour is much more complex and diverse than plant behaviour because the development of a nervous system has facilitated rapid and integrated responses that may include locomotory actions. Slower, chemically based (hormonal) responses are also present (e.g. colour changes in arthropods). There are two main forms of animal behaviour, innate behaviour and learned behaviour, but the distinction is not clear-cut and in higher organisms, the responses often contain components of both.
- **Innate behaviours** are collections of responses which are 'programmed in' to the chemical and nervous systems of organisms such that a given stimulus invariably will produce the same response. These genetically based behavioural patterns evolved over many generations and their main significance lies in their survival value for the species. Innate responses include simple reflex responses, orientation behaviours such as taxes and kineses (Box 5.4), and instinctive behaviours, which may be very complex and include territorial behaviour, courtship, mating, aggression, and social organisation and hierarchies.

Box 5.4

Taxes and kineses: automated responses

A *taxis* is the movement of an entire cell or an organism in response to, and directed by, an external stimulus. A *kinesis* is a non-directional movement in which the rate of movement relates to the intensity of the stimulus but not its direction.

A taxis (tactic response) may be towards (positive, +) or away from (negative, –) the stimulus and taxes are classified according to the nature of the stimulus involved. Phototaxis is the response to light: many unicellular plants are positively phototactic while many animals are negatively phototactic; this response helps them to find shelter by seeking darkened areas. Geotaxis is the response to gravity while chemotaxis is the response to chemicals and includes the positive responses to sexual pheromones and negative ones to noxious substances. Most stream fauna show positive rheotaxis, orientating into the current. Some bacteria are positively magnetotaxic, orientating to magnetic fields.

A kinesis or kinetic response is non-directional but the rate of response is related to the intensity of the stimulus. Thus the tentacles of a sea anemone will respond by increasing their rate of movement if they detect appropriate chemicals or food organisms. The rate of movements of insects may increase with decreasing humidity or increasing temperature. The increased activity level increases the chances of the organism's moving into a more suitable set of environmental conditions although it is not directional.

- **Learned behaviour** requires an adaptive change in individual behaviour as a result of previous experience and depends upon the development of memory. Learning is best known in vertebrates, particularly mammals, and it includes behaviours such as
  - **habituation**, in which repeated exposure to a stimulus results in a diminishing response to that stimulus and
  - **conditioning**, in which organisms learn to produce a conditioned response in association with a particular stimulus: the classic Pavlovian example is the salivation of dogs at the sight of food.
- **Imprinting** involves young animals becoming associated with, and identifying themselves with, another individual, normally the parent. Lorenz demonstrated this in his studies of ducklings and goslings when he showed that the first organism they saw when they hatched was identified as the mother. This led to ducklings and goslings regarding Lorenz as their mother and following him everywhere!
- Another important aspect of behaviour relates to the development of feeding strategies of many organisms (see Section 3.2) as exemplified in the development of optimal foraging by many higher organisms and insects. Social behaviour also represents a fascinating evolutionary step in which the co-operation of individuals of a species results in a higher rate of survival and more offspring than would be possible for individuals alone. It also involves the evolution of altruistic behaviour which benefits another individual at a cost to the performer, a development which at first sight appears to go against the normal tenets of individual behaviour.

The ecological significance of individual behaviour lies in its adaptive potential that is operated upon by selective pressures at the population level to shape future behavioural patterns. How an individual behaves exerts a major influence on its survival and reproductive success. Clearly the distribution and abundance of species is very much influenced by the behavioural ecology of that species, which can be viewed in terms of a cost–benefit analysis. All behaviours have their costs (e.g. energy and mortality costs) and benefits (e.g. improvements in survival and reproductive success). Clearly, organisms do not evaluate costs and benefits but natural selection over many generations moulds behaviour according to these factors.

5.4.3 Dispersal and migration

All organisms are where they are because they have moved there. This is equally true for both mobile and sedentary species and is the consequence of either passive (transport by wind or water) or active (walking, swimming, flying) dispersal mechanisms. *Dispersal* may occur at any life stage in mobile species and could be considered continuous because of locomotory redistribution. Dispersal in sedentary species is usually restricted to the early life stages preceding the final attachment or settlement leading to the adult condition. The results of this spatial reorganisation vary and may lead to clumping (aggregation) in gregarious species and redistribution, shuffling or spreading out and dilution of population density in others. Individual activities thus have population-level consequences.

Temporal variation in conditions and/or resources, induced, for example, by seasonal, diurnal or tidal cycles, may result in the redistribution of individuals (*migration*) or in periods of dormancy within the life cycle of the organism (see Section 5.6). Migration usually refers to movements of groups of individuals, as in the case of locust swarming, tidal movements of shore animals, seasonal movements of wildebeests and intercontinental travels of birds. However, it is the individual that moves and at this level there is no clear distinction between dispersal and migratory behaviour. The key point is that both activities can be

- a means of minimising intraspecific competition,
- a method of colonising new or unoccupied niches or
- a means of responding to spatial variation in conditions and resources (see also Box 5.3).

5.5 Reproduction

Reproduction involves the transmission of genetic material from the parental generation to the progeny, ensuring the passing on of the characteristics (genes), not only of the species, but also of the parental organisms. There are a number of different reproductive strategies, each with its costs and benefits. The fundamental division is into asexual or sexual methods.

5.5.1 Asexual reproduction

Asexual reproduction requires only a single parent and involves only **mitosis**, the cell reproductive process that results in both products of division sharing identical genotypes, subject only to the vagaries of **mutation**. This results in the production of genetically identical offspring or **clones**; genetic change can therefore only result from random, and frequently deleterious, mutations. There are several types of asexual reproduction:

1 **Fission**: found in unicellular organisms (Figure 5.3) and resulting in the production of two (binary fission) or more (multiple fission) identical daughter cells. Multiple fission occurs, for example, in the Sporozoa, a group of parasitic protistans which includes the malarial parasite, *Plasmodium*.

2 **Sporulation** (spore formation): spores are resistant stages produced by bacteria, protistans, all groups of green plants and all groups of fungi. They vary in type and function and are usually produced in very large numbers and in special structures such as sporangia. Being very small and light, they disperse easily.

3 **Budding**: a process in which a new individual develops as an outgrowth (bud) of the parent. It is subsequently released by breaking away from the parent. It occurs in several groups of organisms such as unicellular fungi (yeasts) and coelenterates such as *Hydra*.

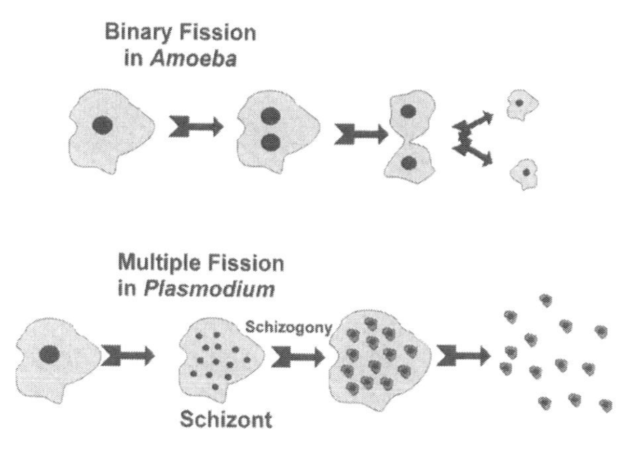

Figure 5.3 ***Types of fission in reproduction***

4 **Fragmentation**: the breaking of an organism into two or more parts, each of which then grows into a new individual. This occurs in some lower animals and the filamentous alga *Spirogyra*.

5 **Vegetative propagation**: a method confined to plants in which a relatively large specialised part of a plant becomes detached and develops into an independent plant: examples are bulbs, corms, rhizomes, stolons and tubers.

5.5.2 Sexual reproduction

This normally requires two parents and involves the fusion of haploid gametes to produce the diploid zygote which then develops into the mature organism. It is an energetically costly process due to the need to produce and release large numbers of gametes. This is particularly true for species using *broadcast fertilisation*, where vast quantities of gametes are required to ensure that a sperm meets an egg but wastage is enormous. However, the most important feature is the mixing of genetic material that occurs during the process of **fertilisation**, the fusing of genetic material from the two genetically different parents. The gametes develop by **meiosis**, a cellular reproductive process which results in a halving of the genetic material to produce the **haploid gametes**. The male gametes (sperm) are typically motile while the female ones (ova) are relatively large and non-motile. The mixing produces genetic variation by genetic **recombination** during fertilisation. Sexual reproduction, therefore, involves the alternation of diploid and haploid phases and the life-form adopted by the organism in each of these phases varies with species. The variability introduced by this mechanism is vital to evolution and adaptation. Table 5.3 compares the asexual and sexual processes.

Table 5.3 ***A comparison of sexual and asexual reproduction processes and their consequences***

Asexual reproduction	*Sexual reproduction*
no gametes involved and therefore no problem of a medium for transmission	requires gamete production and a medium in which they can move
no partner required	partner usually required; therefore, behavioural requirement and cost.
divisions are by mitosis	meiosis involved at some stage in reproduction to prevent chromosome doubling during the production of each new generation
genetic variation results from mutations only; progeny are clones of the parent organism	genetic variation enhanced by a variety of genetic mixing processes (recombination) associated mainly with meiotic divisions
energetically efficient: i.e. low cost	may be energetically wasteful, e.g. broadcast fertilising species where often massive production of gametes is required, most of which do not contribute to the reproductive process
rapid process in most cases, often used to exploit periods of good environmental conditions rapidly	comparatively slow process often used for surviving periods of adverse conditions

There are several variations within the sexual reproductive strategy:

1 Gametes are usually of two structurally distinct types (sperm and ova) but in some primitive forms they are not morphologically distinct and are termed + and – forms instead.

2 Organisms in which the sexes are separate are called *dioecious*. Males and females may be visibly different externally, in which case they are **sexually dimorphic**.

3 Many organisms such as flowers, earthworms, barnacles and garden snails have both sex organs present in the same individual, the *hermaphrodite* condition. This is often regarded as an adaptation to a sessile, slow-moving, low-density or parasitic mode of life. Hermaphrodism would seem to promote self-fertilisation but this raises the problem of inbreeding and its negative genetic consequences (see

Section 6.1). Most hermaphrodite species, therefore, remain obligate or facultative cross-fertilisers and self-fertilisation is relatively unusual except in parasitic forms where finding a mate might be difficult..

4 **Parthenogenesis** is a modified form of sexual reproduction in which the female gamete (egg) develops without any need for fertilisation. Its main advantage is that it mimics asexual reproduction in terms of speed of multiplication. It occurs naturally in both animal and plant kingdoms in two forms – haploid and diploid parthenogenesis. **Haploid parthenogenesis** is common in the social insects (ants, bees and wasps) where meiosis takes place to form haploid gametes. Some eggs are then fertilised to form diploid females while unfertilised eggs develop into fertile haploid males. Aphids use **diploid parthenogenesis**, producing live young which are diploid because of a modified form of meiosis in which the chromosomes fail to separate (total non-disjunction). In plants, parthenogenesis occurs widely in various forms.

5.6 Life cycles and life history strategy

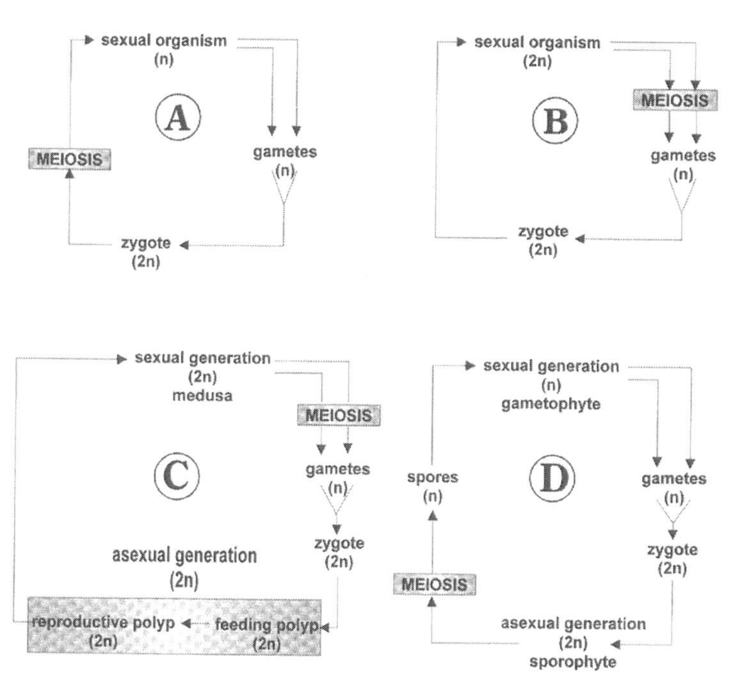

Figure 5.4 ***Four common life cycle patterns. A, Algae such as*** **Chlamydomonas** ***and*** **Spirogyra** ***in which the sexual organism is haploid. B, That found in vertebrates and most animals, where the only haploid cells are the gametes. C,*** **Obelia**, ***in which three different morphological forms occur but all are diploid. D,*** **Laminaria** ***(brown alga) and all land plants where an alternation of haploid and diploid generations occurs***

A life cycle comprises the sequence of stages through which an organism passes from the zygote (fertilised egg) of one generation to the zygote of the next. Life cycles vary in complexity and may involve two or more generations that differ in appearance and reproductive methods; this is an *alternation of generations*. Figure 5.4 shows four common types of life cycle. Land plants and some algae show an alternation between a diploid, spore-producing generation (the *sporophyte generation*) and a haploid, gamete-producing sexual generation (the *gametophyte generation*). Some cnidarians alternate between sexual and asexual generation yet both are diploid, only the gametes being haploid. When different forms of the individuals alternate with each other, as in the cnidarian *Obelia* (Figure 5.5), it is *cyclic polymorphism*.

Parasite life cycles are frequently complex (Figure 5.6) and may involve several stages, each of which may exist in a different, usually very specific, host species.

The life history of an organism comprises its patterns of growth, differentiation and reproduction and reflects the interactions with the environment in which it lives, both biotic and abiotic. The organism's abundance is a reflection of the consequences of these interactions, both positive and negative, and the relative success of reproductive activity. The biology of any organism can be understood, therefore, only by a consideration of its life history and life cycle in relation to its environment. Different organisms allocate differing proportions of their lives to the various functions and each life history is to some extent unique since it is inevitably bound up with the environmental parameters of its habitat. However, a life history is not invariable, despite being fixed within defined limits by the individual's genotype. Plasticity is usually built into any life cycle as the interaction of the genotype with the environment. There are two elements within every life history, therefore:

1 that resulting from the evolutionary process and providing the broad constraints within which the taxon must operate and

2 that resulting from short-term interactions with the immediate environment, modifying responses within the limits set by 1.

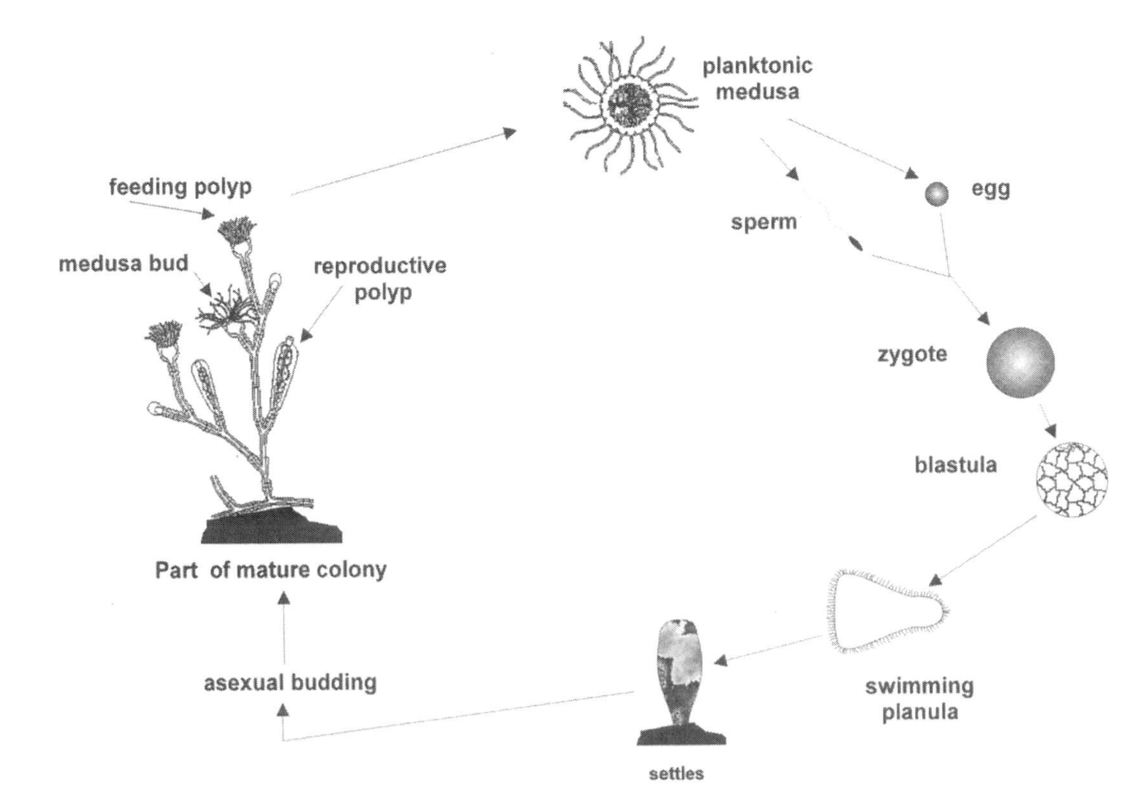

Figure 5.5 ***The life cycle of* Obelia*, a cnidarian with a polymorphic life cycle***

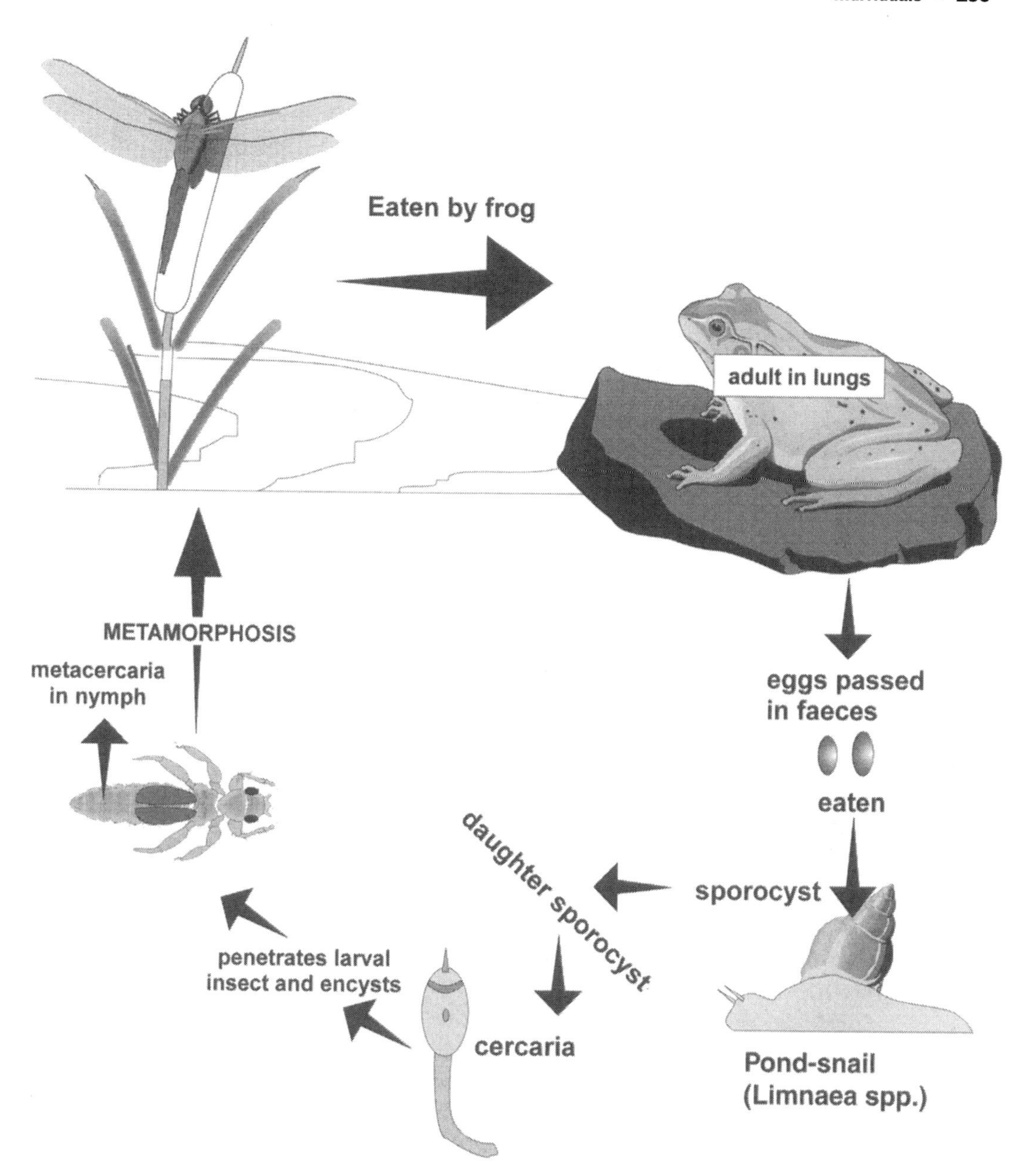

Figure 5.6 ***The life cycle of* Haematoloechus variegatus*, a digenean parasite of amphibian lungs***

What are the main components of life histories? They are:

5.6.1 Size

Size varies with species and from individual to individual within a species; it also varies, of course, with stage of development. Increased size may

- increase competitive ability thus improving reproductive success

- increase success as a predator
- decrease its susceptibility to predation
- enable better homeostatic control as a result of a reduction of the surface:volume ratio.

There are risks, however. Larger organisms may become preferred prey items and they usually need more energy for growth, repair and reproduction. Most studies reveal that intermediate sizes are optimal in terms of metabolic efficiency.

5.6.2 Growth and development rates

Development is the progressive differentiation of morphological and physiological processes, allowing an organism to achieve different functions. Growth and development are quite separate processes in most species. A range of sizes at any particular developmental stage is common, e.g. there is a wide range of sizes in 13-year-old children. Rates of growth and development vary widely and diverse strategies exist. Rapid development typically leads to early reproduction and high rates of population increase. However, many species also include periods of arrested development (**dormancy** and **diapause:** Section 5.6.5) where the life cycle extends through periods of adverse environmental conditions. Longevity is also important in the context of facilitating more reproductive cycles per individual.

5.6.3 Reproduction

Much of life-history variation relates to the details of reproductive methods. Organisms may adopt diverse strategies in relation to the timing and effort of the reproductive period (Figure 5.7). For example, contrast the repeated annual reproductive activity of a tree over decades (*iteroparous* reproduction) with the once in a lifetime reproduction of the salmon or squid (*semelparous* reproduction). Organisms also vary dramatically in their fecundity, a parameter often associated with the method of fertilisation and the degree of parental care. Broadcast fertilisation requires a massive investment in gametes but no aftercare, whilst copulation and parental care (as in birds) allow that investment to be incorporated into the food reserves of a few eggs only. Aftercare of offspring requires the use of resources for protection, suckling and foraging for food for the developing young.

Thus, the reproductive strategy of an organism can have dramatic consequences for both the energy budget of the individual and the abundance and population structure of the species.

5.6.4. Storage mechanisms

Mechanisms for the storage of materials are important for species living in seasonal environments or habitats where resources are irregular in supply. They are

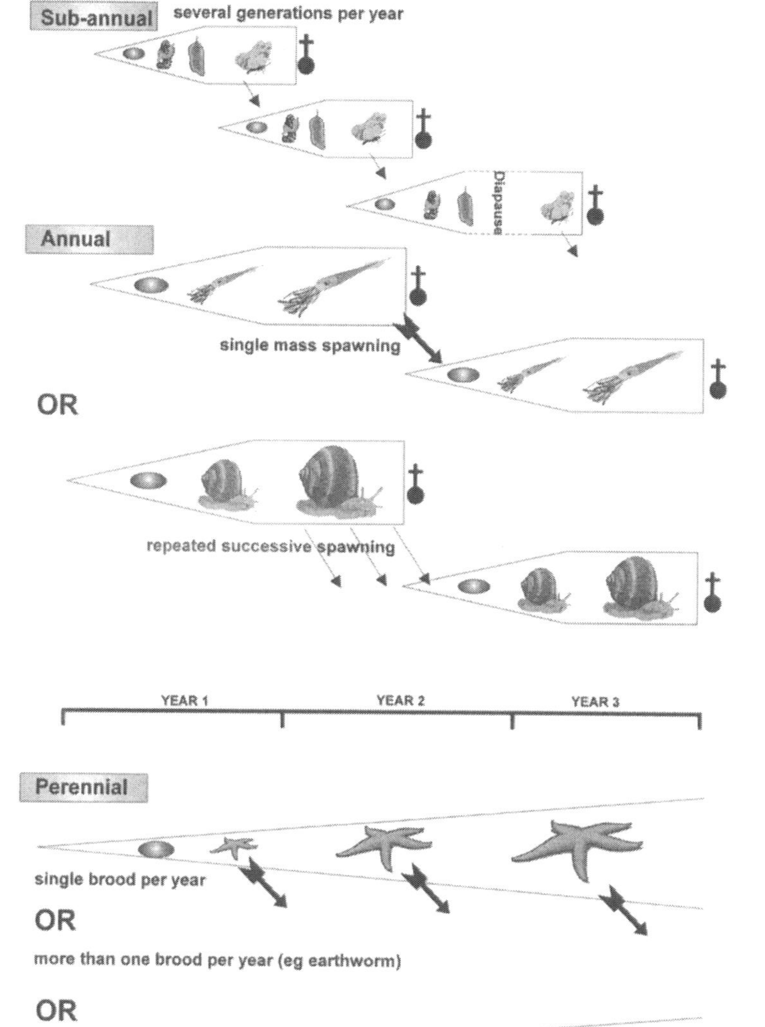

Figure 5.7 ***A summary of the main reproductive strategies in the animal kingdom (adapted from Begon, Harper and Townsend 1996)***

accumulated during periods of abundance and metabolised during periods of shortage for growth, reproduction, defence and maintenance. Stores may be internal in the form of storage chemicals such as fats, starches and glycogen or external in the form of food stores secreted away during good times, e.g. squirrels and their nut stores. Failure to store such materials can lead to large-scale mortality.

5.6.5 Dormancy

The life history of each species and its reproductive strategy fit the organism to its environment while the plasticity built into the types of response allows reaction to unpredictable events. One of the most striking aspects of the life-history modifications associated with periods of adversity is the phenomenon of dormancy.

Dormancy of some kind occurs in almost all types of organisms. Generally it is a physiological state of minimal metabolic activity which allows an organism effectively to avoid periods of poor environmental conditions or resource shortages, or both. The dormancy may be **facultative**, i.e. optional but related directly to prevailing conditions, or **obligate**, in which case it usually becomes associated with a particular life stage, e.g. the **stratification** (freezing) requirement of many plant seeds. Dormancy exists in a variety of forms:

- **Resting spores or buds** are found in a wide diversity of forms with over-wintering or drought/temperature-resistant stages such as bacteria, fungi, plants and lower animals.
- **Diapause** is a form of dormancy that can occur at any specific stage in the life cycle of certain insects. It involves a period of suspended metabolism and complete cessation of growth and development, induced by either temperature or daylength (photoperiod).
- **Hibernation** is a period of inactivity associated with physiological changes which result in a lowering of the metabolic rate. It is an adaptation to conserve energy during periods of environmental extremes found most commonly in mammals living in high temperate regions where winters are severe and protracted.
- **Aestivation** is a period of inactivity associated with hot, dry periods during which the organism remains in a state of torpor with a reduced metabolic rate. Found in fish such as the lungfish and amphibians such as desert frogs.

Dormancy has two important features:

1 One or more life stages must be able to resist extreme or adverse conditions. This is often achieved by the development of a highly resistant external coat.

2 Synchronisation is required between the life stages and the environmental conditions: this is achieved either by a *predictive strategy* or by a *consequential strategy*.

   - A **predictive strategy** results in dormancy occurring in advance of the adverse conditions. It is typical of organisms from predictable, seasonal environments where photoperiod or temperature cues can be used to predict the onset of adverse conditions associated with season. Diapause in insects is a good example of a predictive dormancy and it is unavoidable. Even if conditions do not become adverse, the period of dormancy must take place as it is programmed into the genetics of the species.

   - The **consequential strategy** allows the organism to react directly to current environmental cues and is typical of organisms from unpredictable environments. Its main disadvantage is the potential mortality that might result from a late response. However, its main advantage is that such organisms are able to exploit the environment for as long as possible and only enter a dormant stage if adverse conditions do occur. Predictive dormancy is safer but results in loss of potentially useful periods of habitat exploitation and consequent loss of growth and resources.

5.7 Feeding strategies and mechanisms

Feeding types are broadly described in Section 3.2.1. The majority of living organisms are either photoautotrophic or chemoheterotrophic (Table 5.4).

The key limiting factors for *photoautotrophs* are light, nutrients and water; carbon dioxide is rarely limiting. Individuals are in competition with other species for light

Table 5.4 ***The principal sources of energy and carbon in organisms***

	Carbon source	
Energy source	*Autotrophic (using inorganic carbon dioxide)*	*Heterotrophic (organic carbon sources)*
Photosynthetic using light energy	Photoautotrophs all green plants plus green and purple sulphur bacteria	Photoheterotrophs only a few organisms e.g. purple non-sulphur bacteria
Chemosynthetic using chemical energy	Chemoautotrophs a few bacteria such as *Nitrosomonas* and other nitrogen cycle bacteria	Chemoheterotrophs most bacteria, some parasitic flowering plants, all fungi and animals

and this is an important factor in the evolution of plant form and leaf design. Competition for water and nutrients occurs in terrestrial species through a variety of root designs, adapted to the types of soil in which they occur. In aquatic plankton species, acquisition of light and nutrients is often associated with diurnal movements of the water column, while seaweeds use accessory pigments to assist in the capture of light. Nutrients comprise *macronutrients*, essential elements required in significant amounts for successful growth and nutrition, and *micronutrients* or trace elements, essential elements required in only very small amounts (see also Section 4.2.2.3). Lack of one or more of these nutrients usually results in a deficiency disease. These materials are absorbed from the environment either directly, as in small, usually aquatic organisms, or through specialised absorption systems (e.g. roots in terrestrial plants). The evolution of insectivorous plants is an adaptation to life in nitrogen-poor soils, as is the development of nitrogen-fixing nodules containing symbiotic bacteria in leguminous plants (peas, beans, etc.; see Box 3.7). Mycorrhizae are symbiotic associations between higher plants and a fungus often evolved to assist growth in nutrient-poor habitats.

The other dominant form is *chemoheterotrophic* nutrition, usually termed heterotrophic, only very few bacterial forms being *photoheterotrophs*. Heterotrophs feed on complex organic food, ultimately derived from autotrophs, acquired from the environment by a variety of feeding methods. Digestion and absorption of this complex material, although varying in detail, are remarkably uniform processes (Figure 5.8).

Within the heterotrophic category are a variety of sub-categories:

- **Holozoic nutrition** is the most familiar and involves capture, digestion, absorption, assimilation and egestion, as summarised in Figure 5.8. Within this group are a variety of functional categories including herbivores, omnivores and carnivores (see Section 3.2.1). Organisms feeding on small particles are *microphagous* feeders while those eating large particles are *macrophagous* feeders. Fluid feeders feed on liquids extracted from plants or animals. Organisms feeding on dead or decaying organic material are saprotrophs (saprophytes or saprozoites).

- **Symbiotic nutrition** involves a close relationship between two different organisms and occurs as *mutualism* and *commensalism*. In mutualism, the relationship is beneficial to both partners and is sometimes called *symbiosis*, a surprisingly common relationship. Most reef-forming corals contain symbiotic unicellular algae and ruminant animals have symbiotic bacteria in their gut, providing them with the ability to digest cellulose. In commensalism, the relationship benefits only one partner (the commensal).
- **Parasitic nutrition** involves one organism (the parasite) exploiting another (the host) by living in or on the host. The association benefits only the parasite and the host frequently suffers adversely as a result of the parasite's presence. The most effective parasites are those which do not cause the host to die. They may be:
  - **ectoparasites** such as ticks, fleas and leeches, remaining on the outside of the host or
  - **endoparasites** such as tapeworms, liver flukes and the malarial parasite, *Plasmodium*, which live within the body of the host.

Most parasites are obligate parasites, i.e. they must always live parasitically. Some fungi are facultative parasites since they can become saprotrophic once their host has died.

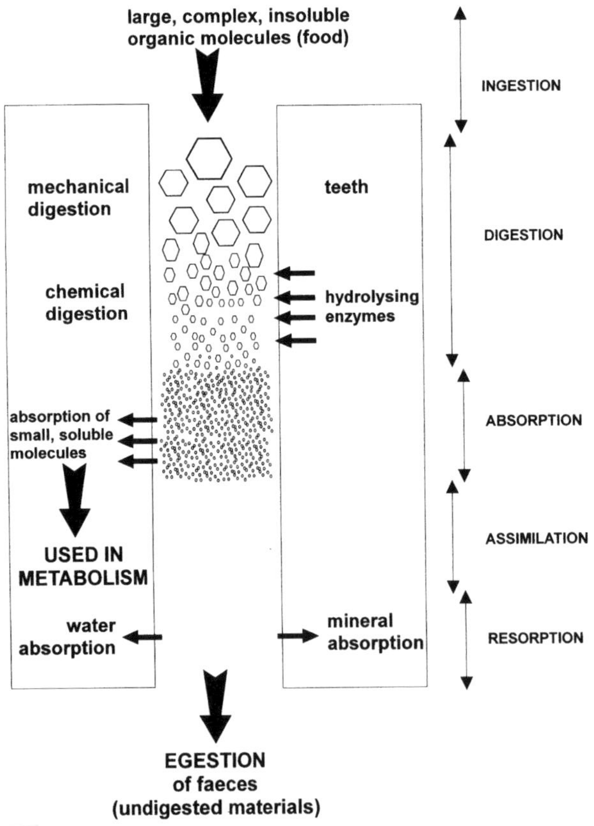

Holozoic feeding is characterised by the evolution of diverse methods of obtaining food material. Sedentary aquatic species require food to come to them and so either create water currents or live in areas where there are natural currents. Mobile species are able to devise a variety of behaviours by which they can chase and capture prey. Feeding types include:

1 Microphagous feeders

- Some use pseudopodia and food vacuoles as in amoeboid protistans, e.g. *Amoeba* spp.
- Others use cilia as in ciliate protists, again using food vacuoles, e.g. *Paramecium* spp.
- Filter feeders use either setose (e.g. *Daphnia* or barnacles) or ciliary (e.g. mussels) mechanisms for generating water currents and for capturing and/or sorting particles.

Figure 5.8 ***The processes involved in heterotrophic nutrition, exemplified by the sequence of activities taking place during passage through an animal's alimentary system***

2 **Macrophagous feeders**

- Scraping and/or boring mechanisms are found in invertebrate grazers such as the gastropod molluscs (snails) and sea urchins. These organisms may have very important ecological impacts, e.g. on rocky shores, because they create and control bare patches of rock (e.g. limpets) or because their grazing controls the growth of algal fronds (e.g. some sea urchins).
- Tentacular feeding is typical of sea anemones, jellyfish and cephalopods (squid and octopus). The tentacles detect and capture prey items, often with the aid of stinging cells (as in cnidarians) or suckers (as in cephalopods).
- A variety of forms that usually ingest their prey whole seize and swallow it. They include many vertebrates such as fish, snakes and many birds. This feeding behaviour is frequently associated with specific adaptations such as recurved and elongated teeth and the development of a large gape (mouth opening).
- Insects such as locusts and mammals such as cattle, elephants and dogs have biting and/or chewing mouthparts. The process of biting and chewing requires mouthparts specifically adapted for cutting and grinding, such as the mandibles of insects and the teeth of mammals.
- Detritus feeders may feed on small particles (e.g. marine polychaetes such as *Amphitrite*) or on larger particles (e.g. earthworms) and use a variety of mechanisms to capture and ingest the food. Earthworms ingest particles directly using a sucking pharynx while aquatic polychaetes often use ciliated tentacle-like structures as conveyor belts to pick up material and transport it to the mouth.

3 **Fluid feeders**

- Sucking species such as the housefly and the butterfly have proboscis tubes or pads used for sucking up fluid such as nectar. Flies secrete enzymes onto potential food to liquefy it for uptake through sponge-like pads.
- Piercing and sucking species include a variety of insects such as mosquitoes and horseflies which suck blood; aphids which suck plant sap; leeches which suck body fluids; and vampire bats which create bleeding wounds with their razor-sharp canine teeth and then suck up the blood.

These diverse feeding strategies and mechanisms are fundamental to the development of an ecosystem, since they form a vital component of the important interrelationships that determine population and community function, as discussed later.

Summary points

- Individuals can usefully be divided into unitary and modular organisms, each type having distinctive ecological features.
- Ecophysiology essentially comprises responses that provide an organism with resistance, tolerance and adaptation to environmental variations.
- Heat balance and metabolism are related to ectothermy and endothermy and their consequences exert diverse effects upon both form and ecology.

- Sedentary and mobile strategies exert major effects on the distribution (dispersal and migration) of species and on their life histories.
- Behavioural mechanisms affect the distribution of mobile individuals and include taxes, kineses and learning behaviours. Sedentary species require pre-settlement strategies to minimise mortality.
- Asexual and sexual reproduction strategies exert major effects on the life cycles and life histories of species. Asexual methods are rapid but lack variation. Sexual methods are slower, often used as overwintering stages, and provide the variation for natural selection to work with.
- Life cycles and life histories reflect both evolutionary and ecological factors. Most have some form of dormancy for the survival of adverse environmental conditions or resource shortages.
- Diverse feeding strategies and mechanisms exist, being mainly photoautotrophic or chemoheterotrophic.

Discussion / Further study

1. What were the reproductive problems associated with the evolution of life from water onto land and how were these problems solved?
2. Why are modular organisms so important in the dynamics of many ecosystems?
3. Select a plant and an animal of your choice from (a) a desert, (b) a high arctic tundra and (c) a tropical rainforest. Investigate the ecophysiology of each species in relation to the environmental conditions it normally experiences.
4. Select one sedentary and one mobile animal species from a particular habitat and contrast their lifestyles and adaptations.
5. Study the life cycle of an aphid. How does it use different forms of reproduction to exploit good conditions optimally and yet survive the adverse winter conditions?
6. Study the life cycle of a tapeworm and the malarial parasite. What fundamental differences and similarities can you identify in their respective parasitic life histories?
7. Discuss the relative importance of different feeding types in (a) a desert, (b) the deep sea floor below 1500 m depth, (c) a tropical grassland such as the Serengeti Plain and (d) the water column of a eutrophic lake.

Further reading

'Life and death in unitary and modular organisms'. M. Begon, J.L. Harper and C.R. Townsend. Ch. 4 in *Ecology* 3rd edition. 1996. Blackwell Science, Oxford.
Useful discussion of the issue of unitary and modular species and their respective ecology.

Migration: Paths through time and space. R.R. Baker. 1982. Hodder and Stoughton, London.
Excellent review of migration and its variations.

'The energetics of lifestyle'. P. Colinvaux. Ch. 6 in *Ecology 2*. 1993. John Wiley and Sons, New York.
Excellent discussion of the ecological significance of ectothermy and endothermy.

Sensory Ecology: How organisms acquire and respond to information. D.B. Dusenbery. 1992. W.H. Freeman and Co., New York.
Modern review of behavioural mechanisms and stimuli including ecological significance.

'Reproductive patterns' and 'Adaptation'. Chs 2 and 5 in *Evolution: A biological and palaeontological approach*, ed. P. Skelton. 1993. Addison Wesley, Wokingham, England.
Useful overview chapters with evolutionary slant.

Introduction to Ecological Biochemistry, 4th edition. J.B. Harborne. 1993. Academic Press, London.
An excellent review of the biochemistry of interactions between animals, plants and the environment, suitable for the more science-based students.

References

Bennett, P.M. and Harvey, P.H. 1987. 'Active and resting metabolism in birds: allometry, phylogeny and ecology', *Journal of Zoology, London* 213, 327–363.

Pough, F.H. 1983. 'Amphibians and reptiles as low-energy systems', in W.P. Aspey and S.I. Lustick (eds) *Behavioural Energetics: The cost of survival in vertebrates*. Ohio State University Press, Columbus.

6 Populations

Key concepts

- **Genetic variation is related to the method/s of reproduction in a species. Sexual reproduction results in genetic mixing.**
- **Genetic variation within a population gene pool is the working material for natural selection.**
- **Inbreeding results in reduced variation within the gene pool while outbreeding increases it.**
- **Many factors influence the genotypes and gene frequencies of populations and small populations are particularly susceptible.**
- **Speciation is the result of mechanisms which isolate populations from gene flow.**
- **Populations of a species occupy a niche but its members are distributed in space (dispersion patterns) and vary in age and size (population structure).**
- **Environments have a finite carrying capacity (K).**
- **Populations are subject to both density-dependent and density-independent controls.**
- **Intraspecific and interspecific relationships are dominated by competition and by predator–prey interactions.**

Populations are groups of individuals of the same species living in a defined geographical area. These individuals share many morphological, physiological and behavioural characters but no two individuals are ever exactly the same, although clonal forms produced by asexual reproduction or parthenogenesis will be extremely similar. Even genetically identical individuals may vary due to differences arising during growth and development. It is this variation that provides the raw material for natural selection. The aim of this chapter is to describe and discuss the characteristics of populations and the parameters, both internal and external, which exert control upon them.

6.1 Ecological genetics

Natural selection is the process by which some members of a population are more likely to survive, reproduce and, therefore, leave offspring than other less 'fit' individuals. Natural selection is the fundamental link between genetics and ecology.

6.1.1 The genetic basis of variation

The basis of heredity, the genetic transmission of a particular characteristic or trait from a parent to an offspring, lies in the structure of DNA (deoxyribonucleic acid) and its organisation into functional units called **genes**. These can be read and transcribed by cellular mechanisms and are ultimately responsible for the growth, development, maturation and reproduction of individuals. Every eukaryotic cell (Section 1.1.2) has a copy of this genetic material, which is effectively the blueprint for the production of that organism, contained within the nucleus of the cell in structures called **chromosomes**. Most cells have two sets of chromosomes (the diploid condition), one set derived from each of the two parents involved in sexual reproduction. This diploid number is maintained during normal cell divisions by the process of mitotic (non-reduction) division. Only during the production of gametes which must contain only one set of chromosomes (the haploid condition) do the pairs of chromosomes separate by the special process of meiotic (reduction) division.

Each gene may have many different forms called **alleles**. If a gene has more than one allele it is polymorphic whilst, if only one form is present, it is monomorphic. These terms can be applied to populations also. A population in which discretely different forms (morphs) of a particular character coexist is said to be polymorphic for that character. Where only one form is present, the population is to be monomorphic for that character. Much of the variation between individuals in a population derives from the fact that many genes are polymorphic.

The complement of genes that an individual possesses is its **genotype** and variations in genotype produce individuals with different appearances or physiologies. The interaction of genotype with the environment produces the individual's characteristics and this outward expression of the genotype is the **phenotype**. Genotypically identical individuals such as clonal plants, grown under different environmental conditions such as rich and poor soils, exhibit different phenotypes. Therefore, some of the variation in a population is attributable to differences in genotypes while some is due to environmentally modified phenotypes.

6.1.2 The sources of variation

The method of reproduction affects the amount of variation occurring in a species population (see Chapter 5). Each new individual will be diploid in the majority of animals and plants and the differences in genotype within a population will depend upon the origins of their chromosomes and their method of duplication.

Asexual reproduction by vegetative propagation (e.g. bulbs), budding, fission or parthenogenesis (Section 5.5.1), results in genetically identical offspring (**clones**) in which the only source of variation is that of random mutation. As a method of introducing variation, mutation tends to be slow and most mutations are deleterious, i.e. cause lethal or sub-lethal changes. A population produced by asexual means alone would be genetically uniform, which is not a disadvantage while it is well suited to the current environmental conditions and these are stable. Its main advantage is that it is a rapid method of reproduction and so the population can exploit the environment maximally. However, when conditions change, perhaps seasonally or because of a disease, this genetic uniformity may result in the current genotype being highly vulnerable. Most plants, many marine invertebrates and some common insects,

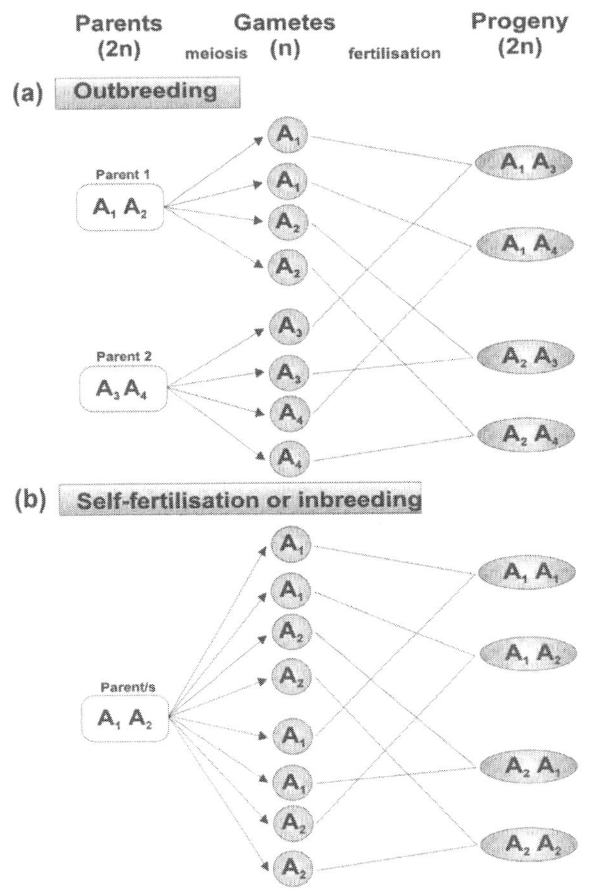

Figure 6.1 ***The effects on genetic variation as expressed by a single polymorphic gene of (a) outbreeding and (b) self-fertilisation or inbreeding***

therefore, have developed a strategy of occasional sexual reproduction within an otherwise asexually based life cycle. This provides an optimal strategy that has the advantages of rapid reproduction, genetic variation and adaptation to environment.

Sexual reproduction is a relatively slow process that involves the production of haploid gametes by meiosis and the subsequent fusion of two gametes (fertilisation) to form a diploid zygote. Both of these processes increase the degree of mixing of genes: meiosis by virtue of the separation mechanism of chromosomes and the associated processes of **recombination** and **linkage**, fertilisation because of the chance processes involved in the meeting of sperm and egg. The gametes may come from the same individual (**self-fertilisation** or inbreeding) or from different individuals (**cross-fertilisation** or out-crossing or outbreeding).

Figure 6.1a illustrates the significance of outbreeding, for which the key consequences are that:

- outbreeding results in offspring which differ from either parent
- there are many different combinations for offspring genotypes
- the two alleles of an inherited gene may be different (the **heterozygous** condition), derived from each genetically different parent.

The largely negative consequences of inbreeding are illustrated in Figure 6.1b. This shows that, because the gametes are from the same organism, half of the offspring contain two copies of the same allele (the **homozygous** condition) and half are heterozygous. Therefore, over long periods, the genetic variation of self-fertilising populations becomes greatly diminished.

The important point is that a high degree of heterozygosity in an individual usually results in it being vigorous and healthy, a phenomenon called hybrid vigour or **heterosis**. Also, if one of the alleles is disadvantageous or fatal, then in the heterozygous condition, this effect may be counteracted by the presence of the other allele. This is the case where the non-damaging gene is dominant and the damaging gene is recessive. A **dominant** gene needs only one of the alleles to be present for its expression while a **recessive** gene requires both alleles to be of the same type for its expression. Occasionally, a recessive but damaging gene present only in the heterozygous condition can confer benefits on the individual. This is the case in sickle-cell anaemia, where the heterozygous condition confers a resistance to malaria (Box 6.1).

Box 6.1

Even lethal alleles may not be all bad

The importance of heterozygotes in maintaining variation illustrated by *sickle-cell disease*. Humans with this disease have irregular or sickle-shaped red blood cells instead of the normal biconcave discs, a feature caused by abnormal haemoglobin which accumulates in the blood cells. These abnormal blood cells cause clogging of blood vessels, a poor circulation, anaemia and/or poor resistance to infections. However, the disease is only expressed when the recessive sickle-cell allele (Hb^S) is present in the homozygous (Hb^S Hb^S)condition: normal alleles are designated Hb^A. When it is present in the heterozygous (Hb^A Hb^S) condition, people have the sickle-cell trait and do not show any signs of the disease except when dehydrated or subjected to mild oxygen deprivation.

In those parts of Africa where malaria is common, infants born with sickle-cell disease die but infants with sickle-cell trait survive better than normal homozygotes. The possession of the sickle-cell allele protects these individuals against malarial infection because the parasite dies when potassium leaks out of the red blood cells as they become sickle-shaped. This results in the sickle-cell allele being maintained in those populations which are also subject to malarial infections; up to 60 per cent of African populations in such areas have this allele. The phenotype of the heterozygote is more fit than either of the homozygous conditions but all three genotypes continue in the population as a result.

Table B6.1 ***The effects of genotype in sickle-cell disease***

Genotype	*Phenotype*	*Consequence*
Hb^AHb^A (homozygous)	normal blood cells	likely death due to malaria
Hb^AHb^S (heterozygous)	sickle-cell trait – blood cells normal except under particular forms of stress	survives due to protection from malaria
Hb^SHb^S (homozygous)	sickle-cell disease with abnormal blood cells	dies due to sickle-cell disease

6.1.3 Patterns of genetic variation

The various genes and genotypes within a single population comprise its **gene pool** and may be only a small proportion of that present in a species over its whole range. The movement of genes between populations is the **gene flow**. How do genetic differences arise between populations of the same species, a process clearly related to the process of speciation itself? The main mechanism is *natural selection* which, since populations are always of restricted size, acts on the inevitably random fluctuation in gene frequencies. Different environments favour particular genes that come to dominate in the population because of the selective pressure.

The four main premises of Darwin's theory of evolution by natural selection are:

1. More individuals are produced than can ever survive.
2. There is, therefore, a struggle for existence because of the disparity between numbers produced and the number that can be supported by the environmental resources.
3. Individuals show variation and those with advantageous features have a greater chance of surviving and reproducing.
4. Since the selected genotypes tend to produce offspring similar to themselves, these genes will become more abundant in the population.

Thus, given certain conditions of variation, inheritance and competition, organisms tend to change by adaptation over time. This gives rise to the concept of *fitness*, defined as *the relative ability of an organism to survive and to leave offspring that can themselves survive and leave offspring*.

However, other factors influence the genotypes and gene frequencies within populations, particularly small ones. The *founder effect* applies to populations which have been established by only a few individuals. It is particularly important on islands and other isolated habitats such as ponds and lakes where the arrival of organisms may be a rare and chance occurrence. Here the founder members will be few in number and their genetic diversity will be low: inevitably, inbreeding will tend to occur with a resulting decrease in heterozygosity. When their progeny are subject to natural selection, the range of possible outcomes is restricted by the limited variety of alleles present in the population, since natural selection can only work with the available gene pool. The results of the same natural selection acting on two small isolated populations may be very different simply because the populations start with very different founder samples from a larger population's pool of genes.

If a population suffers a collapse in numbers due to a disaster such as a disease epidemic, resource shortage or environmental extremes, the small surviving population becomes in effect a founder group. This situation is called a *genetic bottleneck*. The consequences are essentially similar to those of the founder effect, i.e. a low genetic diversity and a tendency for increased homozygosity due to inbreeding. After a population crash, individuals are more likely to mate with their relatives.

Once a population has become established, the degree of isolation it experiences can affect its genetic variability. *Genetic drift* is a random change in the allele frequencies in a population not attributable to the action of any selective process. These differences arise because of biased gamete sampling within small populations. Because gametes are an assortment of haploid genotypes produced by an individual, a small random selection from the range available can form a biased sample of the available genotypes. In large populations, this may not be of any great significance but in small populations, such sampling errors can accumulate over generations producing a genetic drift.

Many populations exhibit a series of phenotypic characters that are clearly adaptations to their particular environment but which are genetically determined rather than environmentally induced. Many plant species have been shown to have different growth forms in different habitats and these different forms have been termed **ecotypes**. They evolve as a result of natural selection acting on genetically isolated populations. When phenotypic variations occur continuously within a population, e.g. along a vertical gradient on a mountainside, any gradual phenotypic change across a population or series of adjacent populations is an *ecocline*.

The pattern of population variations differs with geographical location and the species being considered but essentially it depends upon:

- the degree of isolation – how far apart, spatially or temporally, the populations are
- the amount of genetic exchange between populations – the gene flow
- the type and distribution of environmental characteristics
- the degree of natural selection acting on individuals.

6.1.4 Geographical variation

A good deal of variation may be found between populations in different parts of a species range. Species may show considerable uniformity of phenotype over wide areas where the range is continuous but isolated populations may vary significantly. Where populations are geographically completely separate, they are *allopatric* populations. Virtually all species contain some such isolated populations, especially at the edges of their range, where they are called *peripatric* populations. Where the populations overlap, they are *sympatric* populations.

6.1.5 Mechanisms of speciation (evolution of new species)

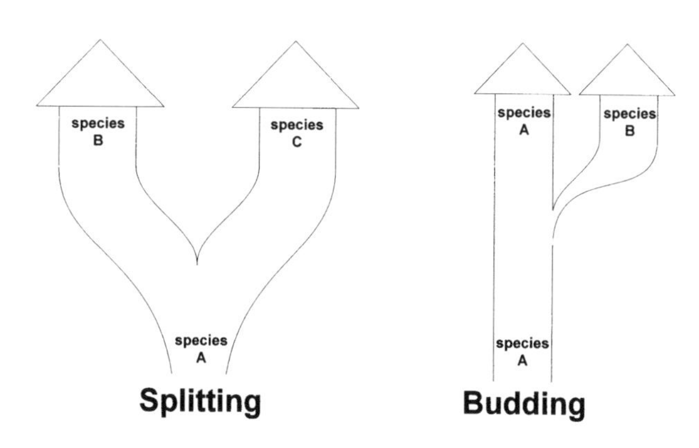

Figure 6.2 ***Diagrammatic representation of speciation by splitting or budding***

Species concepts are considered briefly in Section 1.2.2 but here species will refer to the biological species concept. Consider now how speciation occurs, i.e. how one species divides into two or more separate species during evolution. This can occur in one of two general ways, splitting or budding (Figure 6.2). Splitting involves the gradual separation of one species into two different species and the loss of the original form. Budding is the separation of a new species from the original one which continues unchanged. How does this happen?

Speciation is usually considered in terms of the evolution of *isolating mechanisms* which form barriers to gene flow. Speciation is complete when reproductive barriers are sufficient to prevent gene flow between the new potential species. Isolating mechanisms can be separated into *pre-zygotic* (pre-mating) and *post-zygotic* (post-mating) mechanisms where sexual reproduction and mating occur.

Pre-zygotic mechanisms include:

- **Habitat isolation**, where species are unlikely to meet because of their occupation of different habitats within a common geographic area.
- **Temporal isolation**, where mating or flowering periods are in different seasons, months, or times of day. Closely related species living in the same geographical area are often seasonally isolated.
- **Behavioural isolation**, where the sexual attraction between males or females of different species is reduced or missing due to inappropriate physiology or behaviour. This is one of the most powerful isolation mechanisms in animals where species-specific courtship displays have evolved that utilise a variety of visual, auditory, tactile and olfactory stimuli.
- **Mechanical isolation**, where the mechanical structure of reproductive organs is incompatible between species, thus preventing gamete transfer.
- **Gametic isolation**, in which gametes are incompatible so that fertilisation cannot take place.

Post-zygotic isolation usually refers to some form of hybrid incompatibility such as:

- **Hybrid inviability**, where an egg is fertilised but the embryo does not develop; or the development becomes arrested at some stage; or the hybrid dies before maturity.
- **Hybrid sterility**, where the hybrid survives but fails to mature e.g. the mule.
- **Hybrid breakdown**, where the hybrid is viable and fertile but the viability and fertility of the offspring are reduced.

The prevention of gene flow between two species may be due to one or more of these isolating mechanisms. Remember that such categorisations are often difficult to identify or prove in the real world.

6.2 Population growth and regulation

All populations have characteristic, definable properties such as size, density and dispersion. Although the genetic characteristics of its component individuals and the characteristics of the environment determine a population's characteristics, the population is an operating system with its own distinctive properties. Each population performs a particular role and occupies a particular place in its ecosystem, its *niche* in that system. The members of a population are distributed in space and vary in both age and size, the key parameters in describing a *population structure*. The population size of many species fluctuates considerably while others remain much more constant. Some species are consistently much more abundant than others. What causes a species to be common or rare? Why do population size and structure fluctuate seasonally and annually?

6.2.1 Population size and density

Population size is an important property of any population since it is directly related to a population's potential for survival. Small populations are more susceptible to disturbances and to negative genetic factors (see Section 6.1). The number of individuals of a species per unit area or volume is its *population density*. This parameter can exert a very significant influence on the members of that population and on other species through intraspecific and interspecific competition (see Section 6.2.6). Management of a species usually involves the regulation of population density. Some are managed in order to increase density, e.g. agricultural species and threatened or endangered species, while others are managed to reduce density, e.g. in the control of pest species and disease organisms. Where members of a population differ widely in size, numbers per unit area or volume may not be a useful measure of population size. Biomass (Section 3.2.3), the total mass of individuals, may be a more useful measure.

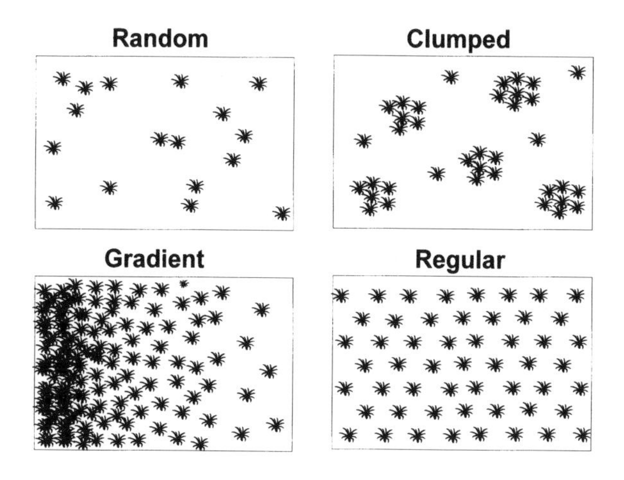

Figure 6.3 ***The four main patterns of distribution of species found in nature***

6.2.2 Population dispersion / distribution

The distribution of a species reflects a variety of aspects of its biology, including patterns of settlement, survival, immigration and emigration. Four general patterns occur (Figure 6.3) although the scale at which the distribution is considered affects the observed distribution pattern:

- **Random distribution** is comparatively rare in nature but usually indicates that few or no environmental factors are influencing the population distribution so that organisms disperse unpredictably.
- **Clumped (contagious) distribution** is found in gregarious species, in species where individuals settle close to their parents and where habitats and/or resources are concentrated in particular areas (patchy).
- **Gradient distributions** are usually associated with environmental gradients such as found on tidal shores, in estuaries and around point sources of pollution.
- **Regular distributions** are comparatively unusual but occur in some plants where competition for resources is intense (e.g. allelopathic species) or in animals where there are social interactions involving territoriality, e.g. birds.

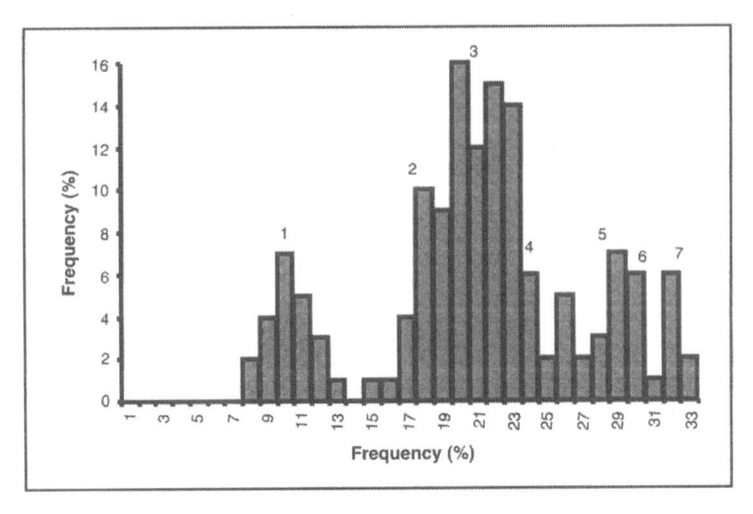

Figure 6.4 ***The age distribution of a population of cockles* (Cerastoderma edule) *from Shetland. The values above the histogram are age in years determined from studies of annual growth rings***

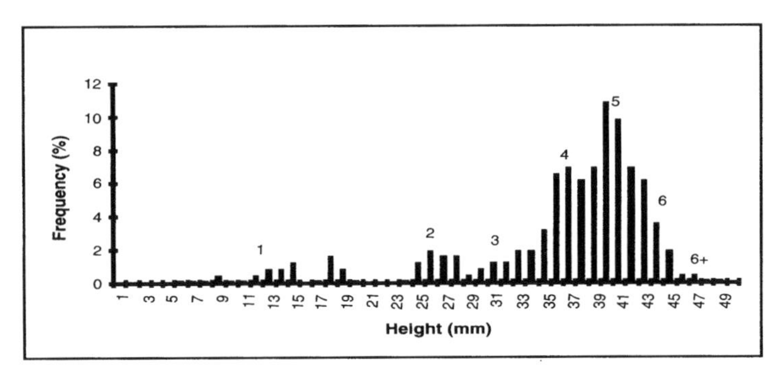

Figure 6.5 ***An example of the dominance of a population of cockles* (Cerastoderma edule) *by two yearclasses. The values above the histogram are age in years determined from studies of annual growth rings***

6.2.3 Size and age structure

Populations comprise individuals of a range of sizes and ages, these two parameters usually being related, smaller individuals being younger than older ones. Age structure, the proportion of individuals in each age group making up its age distribution (Figure 6.4), is determined by the rates of birth and death within the population. If both rates are high, young age groups dominate the population while if both are low, there is a relatively even distribution of age groups. Patterns of age distribution can tell much about the history of birth and death rates in a population. In some organisms such as fish and some bivalve molluscs, a single age group may dominate the population for many years (Figure 6.5).

6.2.4 Population growth

Populations will increase in size whenever birth rates exceed death rates. The rate of population increase (abbreviated as *r*) is expressed as:

$$r = (\textit{birth rate} + \textit{immigration}) - (\textit{death rate} + \textit{emigration})$$

Births, deaths, immigration and emigration are demographic events, whose rates are determined by life-history traits of the species and by the impact of environmental factors, both biotic and abiotic. The number of animals in a population at any given time is:

$$\textit{Nnow} = (\textit{Nstart} + \textit{number born} + \textit{number immigrated}) - (\textit{number died} + \textit{number emigrated})$$

Ecologists frequently construct *life tables* to analyse the pattern of births and deaths in a population (Table 6.1), allowing the calculation of survivorship curves (Figure 6.6) and providing information relevant to wildlife management.

Table 6.1 ***Life table for a cohort of* Poa annua, *a short-lived grass***

Age interval (months)	*Number dying*	*Number alive at beginning*	*Proportion alive*	*Mortality rate*	*Reproduction rate (as number of seeds produced)*
0–3	121	843	1.000	0.144	0
3–6	195	722	0.857	0.270	300
6–9	211	527	0.625	0.400	620
9–12	172	316	0.375	0.544	430
12–15	95	144	0.171	0.625	210
15–18	39	54	0.064	0.722	60
18–21	12	15	0.018	0.800	30
21–24	3	3	0.004	1.000	10
24	–	0	–	–	–

Source: Modified from Purves, Orians and Heller 1995.

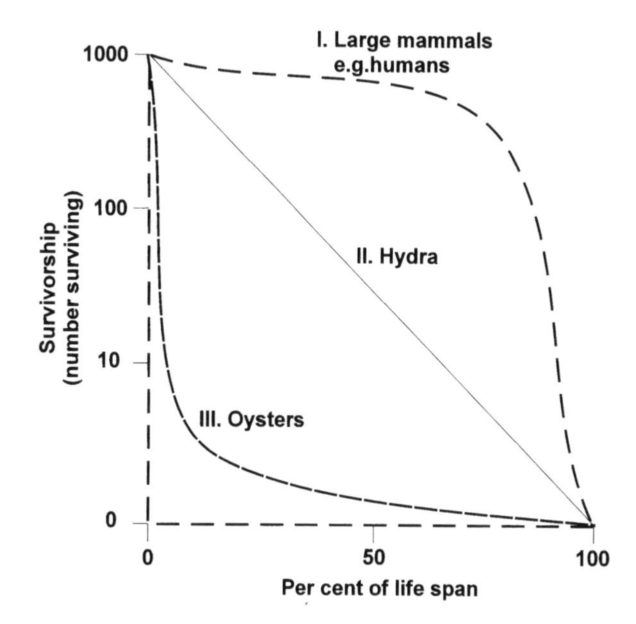

Figure 6.6 ***Three different survivorship curves illustrating different life strategies***

Under conditions of unrestrained growth, a population has the potential for explosive growth because, as the number of individuals increases, the rate of adding new individuals must increase. This results in *exponential growth* (Figure 6.7). However, although populations may grow rapidly for short periods, no real population can maintain exponential growth for very long. This is because, as the population size increases, resource shortages and environmental limitations cause birth rates to fall and death rates to rise. An environment can only support a finite population of a species (see Section 6.2.5), this being called the *carrying capacity* of that environment. This is determined by the availability of resources and by predation, disease and any social interactions such as territorial behaviour. These limitations mean that population growth typically begins exponentially and then flattens out as it approaches the carrying capacity (Figure 6.7), forming an S-shaped growth pattern. This characteristic pattern can be represented mathematically by adding a term which

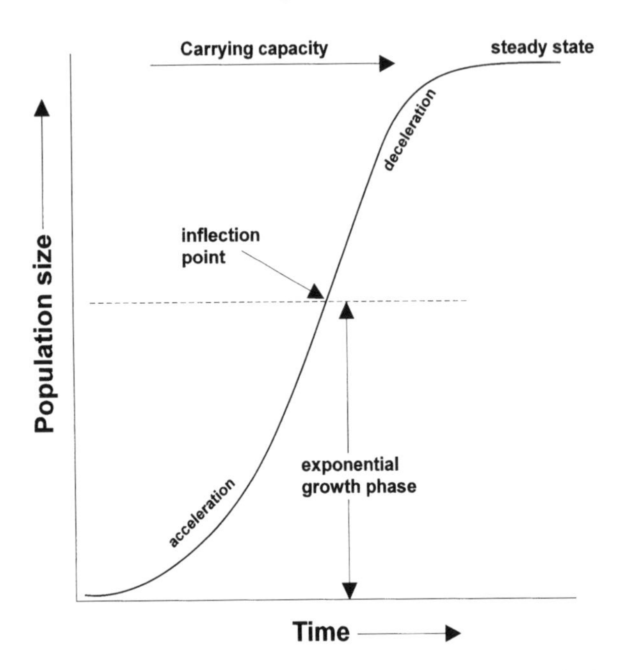

Figure 6.7 ***The S-shaped (logistic) growth curve typical of the growth of e.g. yeast cells in a laboratory system. The density increases exponentially until, at the inflection point, the effects of competition result in a reduced replication rate. Eventually the population size settles to a steady state whose value depends upon the carrying capacity of the system***

slows the population growth as it approaches the carrying capacity: the simplest version is the logistic growth equation

$$\frac{dN}{dt} = r \left(\frac{K - N}{(K)} \right) N$$

where K is the carrying capacity; dN/dt is the rate of change of size of the population; N is the number of individuals; r is the difference between the average per capita birth rate and the average per capita death rate. When conditions are optimal for the population, r has its highest value called r_{max}, the *intrinsic rate of increase*, and this has a characteristic value for each species. Population growth stops when $N = K$ because then $(K - N) = 0$ and so dN/dt becomes zero also. The logistic growth equation contains some important simplifications but it does describe well the hypothesis that population growth in a closed system is subject to density-dependent regulation.

6.2.4.1 Natality (= birth) and mortality (= death) rates

Populations increase because of *natality*, the production of new individuals by birth, hatching, germination or fission. It is usually expressed as the number of organisms born per female per unit time and the rate depends very much on the type of organism. There are many patterns of reproduction, reflecting variations in *generation times*, some species breeding once per year (annual), some several times a year and others continuously. Some organisms such as oysters produce many millions of eggs while others such as birds produce only between one and ten eggs per cycle. This measure of the number of offspring per parent is the *fecundity* of the species and is inversely related to the amount of resources or aftercare provided by the parent.

The rates of *mortality* vary with age, size and sometimes sex. A graph (Figure 6.6), expressed as a *survivorship curve* (the mirror image of death rate), reveals changes taking place over the longevity of that species. Like natality, the rates are influenced by both environmental factors and by the life-history characteristics of the species: of particular importance are competition for resources, adverse environmental conditions and the predator–prey relationships. Patterns of mortality vary markedly between the two hypothetical extremes shown in Figure 6.6, namely:

- where almost all individuals survive for their potential life span and then die almost simultaneously (e.g. salmon and squid) and
- where most die at a very young age but some survive this initial high mortality and the cohort then suffers only a low mortality rate for the rest of its life span (e.g. many bivalves).

Remember that for modular organisms, the concepts of birth rates and death rates are much more complex. They may be applied to the genets, the modules, or both, depending upon the nature of the question being asked.

Life tables are constructed to help determine patterns of births and deaths of a population. They are based on the study of a cohort, a group of individuals born at the same time. The numbers of individuals remaining alive at regular time intervals and the number of offspring produced at those times are measured and expressed as a life table such as Table 6.1. These data allow calculation of birth rates and death rates and the evaluation of patterns of survivorship and population growth.

6.2.4.2 Immigration and emigration

Dispersal of a species is rarely measured in population studies, and it is often assumed that the rates of immigration and emigration are equal or insignificant. However, this is by no means invariably true and in many populations migration is a vital factor in determining and/or regulating abundance. It is difficult to quantify for many organisms and requires identification of individuals before it can be assessed properly. For sedentary species, the dispersal phase is a vital part of the life cycle. However, because it usually takes place as part of the reproductive phase, it has a different significance for the assessment of population parameters.

6.2.5 Population regulation

The activities of populations already discussed, including the factors influencing birth rates, mortality rates, and dispersal, are influenced by both the density of individuals and by environmental changes. The former are *density-dependent* factors and the latter are *density-independent* factors. Competition for resources and accumulation of toxic waste products increase as populations approach their carrying capacities for any particular habitat.

6.2.5.1 Density-dependent controls

Some limiting factors are related to population density, including competition for resources, predation, parasitisation and disease. As a species becomes more abundant, its food supply will be diminished, reducing the amount each individual obtains and resulting in an increase in the mortality rate and/or a decreased birth rate. Predators may be attracted to areas of high prey density and thus the mortality rate will tend to increase. Diseases will also spread more easily in dense populations, again increasing the mortality rate. Density dependence is an important concept in population regulation. If birth rates or death rates, or both, are density dependent, the population responds to changes in density by tending to return to an equilibrium density. If neither is density dependent, however, there is no equilibrium.

6.2.5.2 Density-independent controls

Density-independent factors exert their effects on populations more or less regardless of density and include disturbance factors such as floods, hurricanes, drought, fire and other unpredictable extremes of environmental conditions. Their impact is usually to increase the mortality rate, particularly of smaller individuals, and decrease the reproductive rate. The response of organisms and the effect on population structure and growth depends upon the frequency and severity of the disturbances.

6.2.5.3 r and K strategies

Rates of reproduction vary dramatically from the twenty-minute replication period of some bacteria to that of more than a year for some of the large mammals such as whales. Populations of slow-growing organisms tend to be limited in number by the environment's carrying capacity (K) and are therefore called *K-strategists*; they live in stable, predictable habitats. Populations of species characterised by very rapid growth often followed by sudden and large declines in population size, tend to live in unpredictable and rapidly fluctuating environments and are called *opportunistic species*. They have a high intrinsic rate of increase (r) and are therefore called *r-strategists*. Many species, however, lie somewhere between these two extremes or are able to modify their response according to circumstance.

Generally, r-strategists produce numerous, small offspring that mature rapidly and receive little or no parental care. K-strategists produce only few, relatively large, offspring that mature slowly and often receive parental support; consequently, they are the most susceptible to high death rates and may become extinct relatively easily.

6.2.6 Species interactions

Organisms live within a complex matrix of other organisms of both their own and other species. Interactions between members of the same species are termed *intraspecific* and between species they are *interspecific*; both may be important, although their relative significance varies with circumstance. Such interactions fall into two broad categories: *competition*, where both species may suffer from their interaction for usually limited resources, and *predation*, where one species consumes the other, benefiting one species only. These two categories are, of course, intimately related and predators are often in competition with each other for the same food resources.

6.2.6.1 Competition

Competition between individuals or species is most intense for resources such as space, food and water and obviously increases with density or with reduction in the quantity of a resource, or both. Competition will only be significant if the resource is limiting. When the users significantly reduce the resources, *exploitation competition* results. When one organism, by some means such as behaviour, prevents other individuals from using the resource, it is called *interference competition*. Behaviourally complex animals often interfere directly with each other's activity while other organisms tend to compete by reducing the supply of resources. Competitive interactions may be very complex and lead to short-term fluctuations in abundance and distribution of a species and to long-term evolutionary adaptations providing a species with improved competitive ability within its particular niche.

The negative effects of competition result in features such as reduced growth or fecundity, exclusion from a habitat, or mortality. Intraspecific competition may be severe since individuals of one species have essentially similar requirements and therefore compete most directly. However, many species overlap significantly in their resource requirements, leading to interspecific competition which may also be intense. It is particularly strong between plant species because most require the same mineral nutrients and all utilise sunlight. Gause (1934) demonstrated that when two directly competing species within a simple, uniform, closed system (micro-organisms within a test tube) interact, one species will eventually completely exclude the other: which species wins depends upon the conditions within the system. This process is called *competitive exclusion* and is typical of restricted environments. However, in patchy environments where each species can find local conditions that favour its survival and propagation, a species may persist even when its population densities are reduced by competition in some parts of its distribution.

6.2.6.2 Predator–prey interactions

Predator–prey interactions, the interactions between food resources and their consumers, exert very significant controls on populations. The types of predators are considered in Section 5.7 and include herbivores, carnivores, parasites and parasitoids, suspension and filter feeders. It also includes commensalism, mutualism and parasitism, forms of interaction which are too specialised to include here.

When a predator eats a prey item, it reduces the prey population. If the predators can find enough prey to eat, they are sustained, but once the prey density is reduced to below a threshold, the predators themselves will be reduced by starvation or emigration. This reduces the predation pressure on the prey species so that it will recover and increase again, and so the cycle repeats. Thus predator–prey interactions would be expected to cause population fluctuations of both components, producing alternating oscillations in populations of predator and prey species (Figure 6.8). All other things being equal, this should result in an equilibrium population size of both predator and prey populations. However, in closed system experiments (e.g.

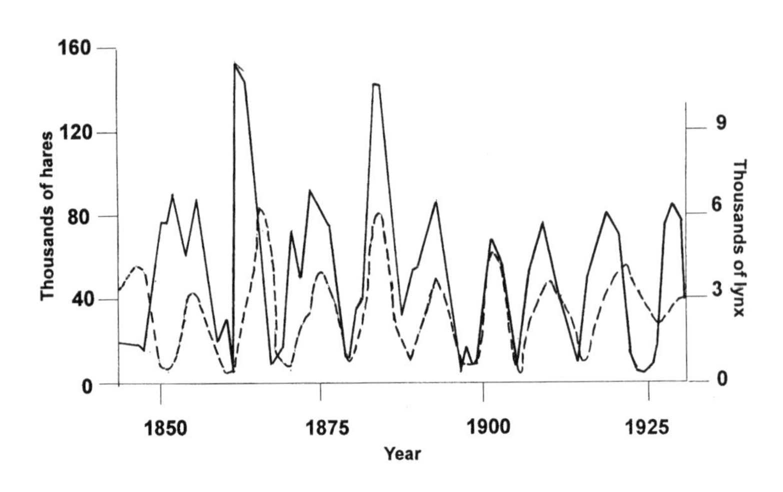

Figure 6.8 ***Oscillations in a predator (lynx) – prey (snowshoe hare) system in Canada, based upon long-term data from the Hudson's Bay Company***
— *snowshoe hare,* Lepus americanus
---- *lynx,* Lynx canadensis

Gause 1934), where a predator protozoan *Didinium* interacted with a limited population of the prey, *Paramecium*, the predator first ate all the prey and then the predator itself died from starvation. That is, no equilibrium resulted. The nature of such interactions in an open system depend upon many factors such as relative sizes, reproduction rates, growth rates, etc. and there is much debate about the significance of predator–prey interactions, which vary considerably between different species and environments. Population cycles are characteristic, however, of some species of small mammals such as lemmings and these are sometimes related clearly to predator abundance.

The dynamics of predator–prey interactions form the basis for the development of biological control methods where the optimal situation is to reduce, but not eliminate, the prey species (the pest). This enables the maintenance of the predator population and ensures that it does not become extinct. In the same way, less virulent strains of disease-causing organisms are favoured by natural selection and will survive while virulent strains will eliminate their hosts and thus also eliminate their source of maintenance (Box 6.2).

The complex interactions between predators and their prey are important for the maintenance of community diversity. This is because, by controlling the relative abundance and distribution of some species, they provide opportunities for other species by reducing competitive exclusion. They also result in long-term evolutionary change, since predation is an important selective factor, prey species evolving a rich variety of responses to potential predation which make them more difficult to capture and eat. This includes development of spines, hairs and bristles, toxic or noxious chemicals, mimicry (Box 6.3) and behavioural modifications. In their turn, predators also evolve to become more effective in overcoming prey defences.

Box 6.2

Rabbits and myxomatosis: biological control at work

The European rabbit, *Oryctolagus cuniculus*, which had been introduced into Australia had reached the status of a pest of grazing lands by the 1950s. In an attempt to control it using a disease-based form of biological control, the myxoma virus from a South American rabbit, *Sylvilagus brasiliensis*, was introduced to the Australian populations. This virus caused only a mild disease in the South American rabbit but was usually lethal in the European rabbit.

The consequence of this introduction was a rapid and massive (up to 90 per cent) reduction in the rabbit populations. The virus was then introduced into the United Kingdom and France, where it caused similar mortalities. However, studies over a long period of time revealed several important features of such viral diseases. Perhaps the three most significant were:

- The virulence of the virus declined markedly over the next decade. Initial mortalities were as high as 99 per cent but fell to 90 per cent within a year and continued to decline until its effectiveness was reduced by more than 50 per cent within eight years.
- The rabbit populations became more resistant to the virus anyway. Selection for resistant strains in the wild populations was inevitable because of the lethal nature of the infection.
- The initial vectors of the disease were mosquitoes but because the virus killed the rabbits so rapidly, there was only a short period during which the vector could transmit the disease from infected rabbits. The result was that the virus was not maintained very well within the rabbit population.

Similar patterns of change were observed in the United Kingdom and French rabbit populations. Biological control is always subject to very complex variables and its use should be carefully considered.

Interestingly, in late 1995, a deadly rabbit calcivirus which causes death by asphyxiation within two days escaped from a research facility on Wardang Island in South Australia. It is proving to be particularly virulent, millions of rabbits dying within the first few months (Anderson 1995). It is spreading much more rapidly than expected and its long-term impact remains uncertain.

Summary points

- Genetic variation is related to the method/s of reproduction in a species. Sexual reproduction results in genetic mixing while asexual processes produce clonal forms in which mutation offers the only source of genetic variation.
- Genetic variation within a population gene pool is the working material for natural selection.
- Inbreeding reduces genetic variation and vigour while outbreeding and increased heterozygosity increase it.
- Many factors influence the genotypes and gene frequencies of populations, particularly small ones. Important factors are the founder effect, genetic bottlenecks and genetic drift.
- Speciation is the result of a range of mechanisms which isolate populations from gene flow although they may still have overlapping distributions These mechanisms fall into two categories, pre-mating and post-mating (hybrid incompatibility).

Box 6.3

Mimicry: cunning disguises

Many organisms try to blend into their background, mimicking foliage or background. This is **crypsis**, the development of cryptic body forms, while behaviours such as catalepsis (use of a frozen posture; playing dead) are also forms of mimicry. Some animals mimic other animals and there are three main strategies, all of which require the involvement of three components: the model to be mimicked; the mimic; and the predator/prey that is to be fooled (duped).

- **Mullerian mimicry** is where unpalatable animals evolve to look the same (evolutionary convergence) as other unpalatable species. This reinforces the relationship between the design and the unpalatability feature and provides a group defence. This is seen in the sharing of design and coloration of many tropical butterflies and in many kinds of bees and wasps. Species that give warnings are termed **aposematic** and the warnings may consist of coloration (intense colours), patterns (black and yellow stripes in insects), smells (skunks) or sounds (rattlesnake rattles).
- **Batesian mimicry** is the mimicry by a palatable or harmless species of an unpalatable or dangerous species. Innocuous hoverflies (Syrphidae) are striped black and yellow (warning coloration) to resemble stinging bees and wasps. Similarly, pseudocoral snakes mimic the poisonous coral snakes.
- **Aggressive mimicry** is the term given to mimicry not used to escape predation but used to facilitate predation by the mimic. This is usually a mimicking of the background and is really a form of crypsis. Good examples are the mimicking of flowers and twigs by preying mantises and the mimicking of rocky substrates by angler fish and spider crabs. Female fireflies of the genus *Photuris* mimic the flashing light patterns of other species to lure the males of those species within range to eat them.

- Populations of a species will occupy a niche but its members are distributed in space (dispersion patterns) and will vary in age and size (population structure).
- Environments have a finite carrying capacity (K) determined by factors such as resources, predation, disease and social interactions.
- Populations are subject to both density-dependent and density-independent controls.
- Intraspecific and interspecific relationships are dominated by competition and by predator–prey interactions.

Discussion / Further study

1. What is the Hardy–Weinberg equation? Consider its significance for population genetics.
2. What are ecotypes? Describe three examples from the literature.
3. How do organisms colonise new islands? What is the significance of such events for speciation?
4. Why is a truly random distribution of individuals in a population so rare? Why are clumped patterns of distribution the most common?

5 What is the potential significance of emigration and immigration for a population? Select a plant and an animal species and discuss for each.

6 How do sedentary species such as plants and barnacles reduce the problem of intraspecific competition?

7 Many organisms we wish to conserve have low reproductive rates and are long-lived while most pest populations are short-lived but with very high reproductive rates. How do these characteristics affect management strategies for such organisms?

8 Which is best able to adapt to environmental change, a human being or an insect? Explain your reasoning and describe how such species are able to adapt to any such change.

Further reading

Games of Life: Explorations in ecology, evolution and behaviour. K. Sigmund.1993. Oxford University Press, Oxford.
A beautifully written survey of ecological and evolutionary ideas.

Population Ecology: A unified study of animals and plants, M. Begon, and M. Mortimer. 1986. Blackwell Scientific Publications, Oxford.
An excellent introduction to population ecology with well-illustrated case studies.

Interactions: Part 2 in *Ecology*, 3rd edition. M. Begon, J.L. Harper, and C.R. Townsend. 1996. Blackwell Science, Oxford.
A broadly defined treatment of population biology, superbly presented.

'The harvesting of interacting species in a natural ecosystem'. J.R. Beddington and R.M. May. November 1982. *Scientific American* 247, 62–69.
A study of whales and other animals feeding on krill populations as an example of the problems of using a biological resource without destroying it.

Living in the Environment, 8th edition. G.T. Miller. 1994. Wadsworth Publishing Co., Belmont, CA.
Chapter 6 for population biology generally and Chapter 8 for human population dynamics and regulation. Well-written and illustrated accounts particularly suited to environmental science students.

Conservation Biology in Theory and Practice. G. Caughley and A. Gunn. 1996. Blackwell Science, Oxford.
Chapters 5 and 6 deal concisely with population factors as they influence conservation matters.

Population growth and Earth's human carrying capacity.' J.E. Cohen. 21 July 1995. *Science* 269, 341–346.
Considers the question of how many people the planet Earth can support.

References

Anderson, I. 9 December 1995. 'Runaway rabbit virus kills millions'. *New Scientist* 269, 4.

Gause, G.F. 1934. *The Struggle for Existence.* Williams and Wilkins, Baltimore, MD.

Purves, W.K., Orians, G.H. and Heller, H.C. 1995. *Life: The science of biology.* Sinauer Associates Inc., Sunderland, MA.

7 Communities and ecosystems

Key concepts

- **A community is a dynamic and interactive collection of populations.**
- **An ecosystem comprises the biological community together with its physical environment.**
- **Communities are defined either by the habitat or by the dominant life-forms and occupy a variety of scales.**
- **Almost all populations and communities are distributed patchily.**
- **Community interactions are founded upon trophic relationships and/or predator–prey relationships.**
- **Primary and secondary productivity varies markedly between ecosystems.**
- **Changes in composition and form of community vegetation follow distinct patterns known as ecological successions, usually ending with a relatively stable climax community.**
- **Biodiversity includes habitat, species and genetic components and is under threat from human activities.**
- **Stability of communities and ecosystems is a function of three properties: persistence, constancy and resilience.**
- **Biogeographic distributions are the result of both historical (evolutionary) and ecological factors.**

The previous chapter shows how interactions at population level influence population dynamics, with population characteristics being shaped by environmental pressures, both biotic and abiotic. An ecological **community** is a dynamic collection of species populations occurring together in space and time within some common, defined habitat or environment; they are integrated or interact so as to influence other component members. The concept originated from the study of plant aggregations (phytosociology) but now applies to all organisms. Most communities include mixtures of members from most kingdoms. Groups of populations forming a community are linked together by a complex range of interactions, directly or indirectly linking all its members together in a web. This web is very much based upon competitive and predator–prey relationships but many of the linkages are very subtle or ephemeral. Examples are plants requiring animals for pollination or dispersal, and facultative or obligate mutualism, where the performance of one species is enhanced by, or dependent upon, the presence of another organism. Such is the complexity of community dynamics that our understanding of it is still fairly limited. However, its importance for the prediction of environmental impacts and for the development of environmental conservation and management programmes cannot be overstated.

The study of community ecology has largely been descriptive because of the inherent difficulty of observing, measuring and experimenting within such complex systems.

Communities are defined either by the environment or the habitat in which they occur or by the dominant species in the association, e.g. a lake community and a grassland community. Communities can be of any size, ranging from the microbial communities within the rumen of herbivorous mammals to the vast expanses of tropical rainforests. Why do the same groupings of organisms occur again and again? The answer lies with the concept of the ecological niche and the feeding relationships within the system (see Section 3.2 and Chapter 2).

7.1 Habitats and niches

Any habitat is best considered as a collection of niches. The difference between a habitat and a niche is a fundamental one for ecology. A *habitat* is the physical or geographical place where an organism lives and can itself be subdivided into *microhabitats*, small, specialised habitats containing a limited range of organisms (Figure 7.1). The carrying capacity of a particular habitat, the maximum number of supportable organisms, depends upon the availability of resources within that habitat.

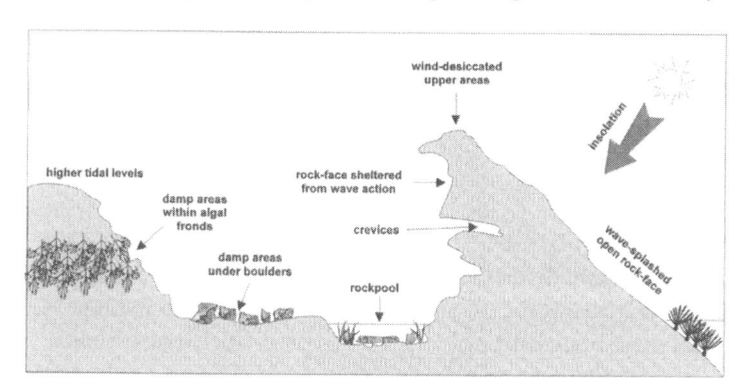

Figure 7.1 ***The diversity of microhabitats found in a typical rocky-shore environment***

A *niche*, however, is the unique position occupied by a particular species in a community. It comprises the physical, chemical, biological, spatial and temporal factors required for the survival of that species and thus determining its distribution and dynamics, i.e. its ecological role. Providing a quantitative description of a niche is both difficult and complex since it is in effect a multidimensional concept (Box 4.1). Hutchinson (1958) defined a species niche as an n-dimensional space (hypervolume) with n – 1 different environmental variables (see Box 4.1). Each species has its own unique niche since, if two organisms occupied exactly the same niche, one would out-compete the other, forcing it to extinction. However, two species may occupy overlapping niches, a feature common in nature and the basis of competition. The concept that no two species can coexist if they occupy exactly the same niche is *Gause's competitive exclusion principle* as described earlier (Section 6.2.6.1). So how can species coexist? The answer is simply because they occupy, often subtly, different but overlapping niches within the system. Hutchinson (1957) also divided niches into two types. An organism's *fundamental niche* is the niche it could occupy in the absence of competitors and predators. Its *realised niche* is the one actually occupied within a particular community after all possible interactions have been exercised.

The consequence of overlapping niches and the resultant competition for resources is the evolution of *resource partitioning*, i.e. the dividing up of the use of each resource by species specialisation and adaptation. This allows the exploitation of different components of the resource by different organisms. It is frequently associated with character evolution in morphologically similar species where competition and natural selection result in the modification of these morphologies in accordance with resource partitioning. The evolution of different beak types in the Galapagos finches as described by Darwin is a good example. A number of communities, such as the large grazing mammals of the Serengeti Plain in Tanzania, also illustrate resource partitioning. The concept of carrying capacity (Section 6.2.4) is clearly related, therefore, to the availability and distribution of different niches.

7.2 Patchiness

The most conspicuous feature of the local distribution patterns of populations and communities is their patchy nature. The four main types of distribution patterns (Figure 6.3) are:

- random, which is rare in nature
- aggregated, the most common form, also known as patchy, clumped or contagious distribution
- gradient, which is common wherever physico-chemical gradients are well defined, e.g. tidal shores
- regular, which is uncommon but found where adaptations to intense spatial competition have evolved, e.g. creosote plants.

The exact distribution a species exhibits within a community depends upon the scale at which it is studied. However, patchy distribution is the most common at the level of population or community. The reasons for patchiness are many and varied but it is interesting to consider that most organisms are absent from most environments most of the time, reflecting the aggregated nature of most distributions. The environment is rarely uniform enough for the development of species exclusion and often there is not time for a competitively dominant species to exclude completely a competitively inferior species before environmental conditions change. Thus competitive interactions rarely proceed to the theoretical state of equilibrium and competitive status may be in a state of almost continuous flux, allowing the development of species-rich communities. This leads to the development of patch-dynamics models, an important class of models that views communities as consisting of a number of patches which can be colonised randomly by individuals of a number of species. An important component of such systems is disturbance, which can act as a resetting mechanism (see succession, Section 7.4). The main reasons for the development of patchy distributions include:

- the spatial complexity of the environment, where conditions and resources are rarely distributed evenly
- the temporal variability of the environment, where the distribution of conditions and resources usually vary with time as well as space

- temporal and spatial changes in the dynamics of grazing and predation pressures on component species, usually associated with foraging strategies
- temporal and spatial changes in intraspecific and interspecific competition
- events in time that remove organisms and open up space for colonisation by the same or different species.

These interactions are very complex and it would be surprising if organisms were not patchily distributed. Patchiness is generated by any physical or biological activity that creates either an open space for recolonisation, a new substrate or niche, or a change in the prevailing conditions or dynamic interactions acting upon an area. Inevitably, prey species are reduced or eliminated faster in habitats where they are more vulnerable, leading to very rapid changes in relative abundance and distribution. Some communities will contain competitively dominant species forming *dominance-controlled* communities typically showing predictable successional sequences (Section 7.4). Others contain species with similar competitive and colonising potential and are *founder-controlled* communities, e.g. rocky shores where a competitive lottery takes place between species when vacant spaces occur (Box 7.1).

Patchiness is of considerable significance in field ecology since it imposes a need for appropriate sampling methods. Such methods must take both the nature and scale of patchiness into account. This may become particularly important when we are considering conservation matters such as habitat fragmentation, where patch size and edge effects may become important considerations for management.

7.3 Trophic interactions and productivity

The fundamental concepts of trophic structure and interactions are described in Section 3.2. The community is a system for the exchange of energy and organic matter between three basic components: the primary producers, the consumers and the decomposers. Primary producers are the autotrophic organisms which synthesise 'food' from basic raw materials. The consumers comprise the heterotrophic members of the community that feed upon the producers (primary consumers) or upon each other (secondary consumers).

The decomposers are a distinctive and vital part of any community and comprise saprophytic or saprozoic organisms feeding on dead and decaying materials derived from the two other groups. Such a classification is, of course, oversimplified since some organisms will fit into more than one of these categories. While available matter is recycled, energy passes linearly through the trophic structure and is not recycled, being finally dissipated as heat. Thus, the Eltonian pyramids described in Section 3.2 are widely applicable. The consequence of the major losses of energy during trophic transfers (gross ecological efficiency typically of between 7 and 14 per cent) is that in most terrestrial communities there are rarely more than four or five trophic levels.

It is also important to recognise that communities differ both in the number of trophic levels they contain and in the relative importance of each trophic class within them. This reflects differences in the dynamics of each community. In particular, the amount

Box 7.1

Plankton patchiness

Despite the apparent horizontal uniformity of the pelagic environment, a characteristic feature of **plankton** populations (organisms that are incapable of swimming from current system to current system) is that they tend to occur in a patchy or non-random manner (spatial and temporal heterogeneity). This feature occurs on a variety of spatial scales from a few metres to hundreds of kilometres, and on a temporal scale from daily to annually. It also happens vertically within the water column. Such non-random distributions are usually due to physical, chemical or biological events.

Physical determinants of patchiness operate on all types of plankton. Various types of horizontal and turbulent mixing, together with large-scale advection or movement of water masses such as upwelling, gyres and eddies, can result in aggregation or dispersion of plankton. Any factor that affects the density of water can result in the separation of water masses from each other, as shown by the occurrence of coastal fronts, thermal stratification (thermoclines) and salinity gradients. This separates waters with different physical, chemical and biological characteristics from each other and facilitates the independent development of the biota within these patches.

Biological factors causing patchiness usually act on small scales and involve a variety of biological processes. The primary ones, however, are those associated with

- the locomotory and sensory behaviours of planktonic species,
- reproductive activities that tend to result in gametes and larval stages being concentrated within the water body and
- grazing and predator–prey interactions and their consequences.

Planktonic organisms may respond to local physico-chemical variations by differential locomotory activity. Thus planktonic copepods may migrate vertically over several hundreds of metres every day in response to the light–dark cycle. The reproductive cycles of both planktonic and benthic species frequently result in major concentrations of gametes or larvae in the water column. A good example is the massive release of coral gametes which usually occurs on one particular night each year, determined by the lunar cycle. Small-scale patches of phytoplankton can also result from the dynamic nature of the interactions between phytoplankton reproduction rates and variations in zooplankton grazing pressure. A patch of phytoplankton offers rich feeding to herbivorous zooplankton that tend to concentrate on this patch, thus relieving grazing pressure on adjacent areas and allowing the phytoplankton to multiply there under reduced grazing pressure.

Some organisms can produce toxic reactions in other organisms when they are in concentrated patches and these toxic algal blooms occur in both freshwater and marine systems (see Box 3.6).

of energy available to different communities (net primary production – NPP; Section 3.1.2) varies enormously (Figure 7.2). This has profound implications for the structure (including diversity) and mode of function of each community.

Within a community, the identification of functional clusters of species populations that interact more strongly amongst themselves than with other component species led to the concept of *guilds*, groups of species that exploit the same environmental resource base in a similar way. This groups together organisms with overlapping niche

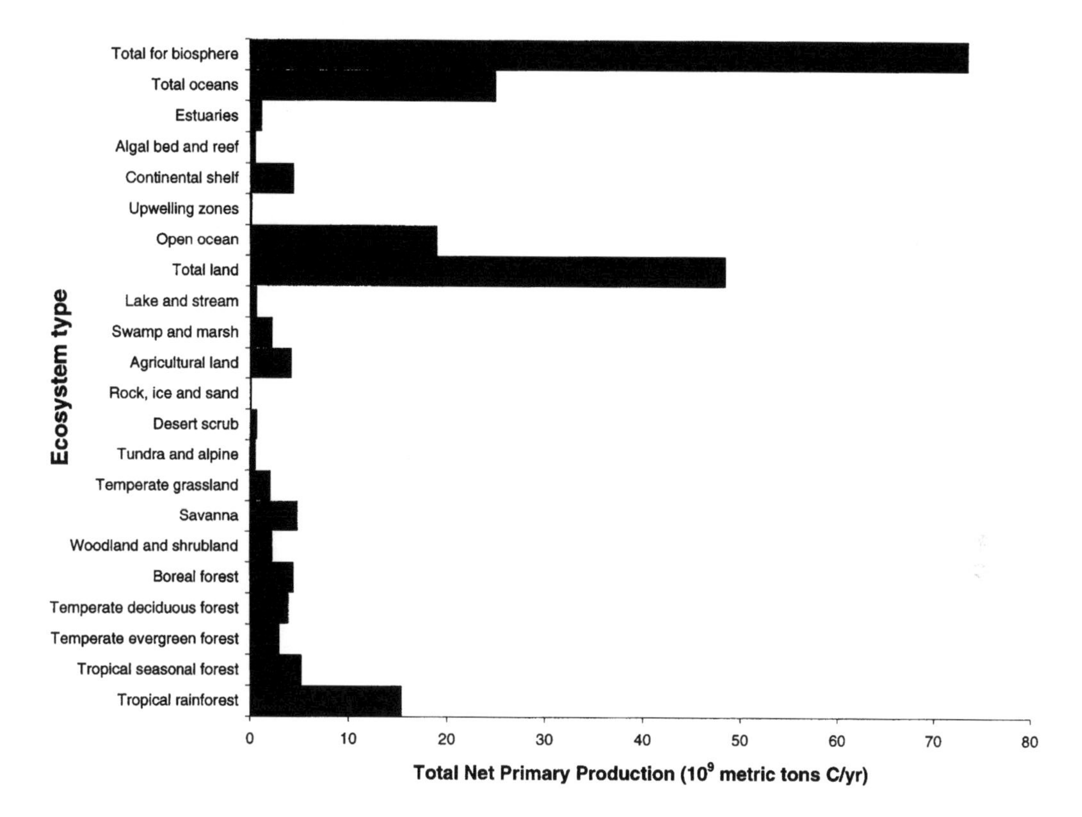

Figure 7.2 ***Variation of total net primary production in the world's major types of ecosystem***

requirements, which therefore compete strongly with each other. Thus within the insectivorous birds one can identify subgroups (guilds) such as ground feeders, flycatchers, etc.

Some communities are structured in such a way that a single species may have a critical role for that community. Such species are *keystone* species and changes in their abundance cause major changes in the community structure. The limpet *Patella vulgata* is a good example, exerting control of wave-splashed rocky midshore communities around the British Isles. The grazing activity of the limpets maintains a diverse community by maintaining patchiness and their exclusion or removal results in dense stands of opportunist species until natural mortality allows the limpets to re-establish themselves.

Dominant species may also exert a powerful control over the occurrence of other species. They are recognised by their numerical or biomass abundance, e.g. the major plant species that dominate large areas of climax woodlands. Dominance can arise by three main processes:

- by rapid colonisation of a new resource by opportunistic (r-strategists: see Section 6.2.5.3) species
- by a species specialising on one part of a resource complex that is widely distributed and abundant

- by a species becoming a generalist so that it is capable of utilising a wide variety of resources.

Dominance is an important part of community organisation and dominant species typically structure other parts of the community and may well affect both diversity and stability of that community (Section 7.5).

7.3.1 Measuring productivity

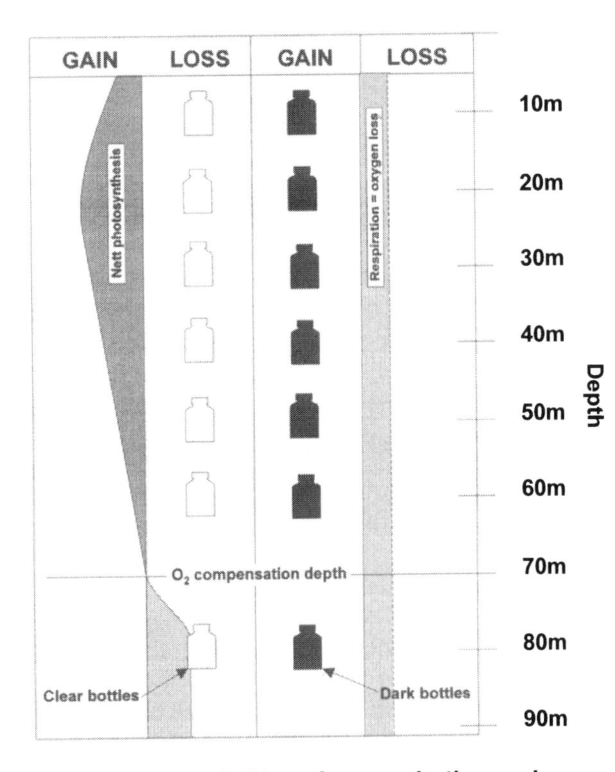

Figure 7.3 ***Phytoplankton primary production can be assayed biologically by comparing the oxygen levels in clear (light) bottles and dark bottles containing phytoplankton at different depths. The change in oxygen content provides a measure of the net production or loss of biomass at that depth. Oxygen produced by photosynthesis is also used for respiration: only when there is a net oxygen gain (i.e. when photosynthetic production exceeds respiratory losses) can the plant persist. The depth at which the two processes are balanced is the oxygen compensation depth. The dark bottle measures respiratory losses only while the clear bottle measures net oxygen changes: by adding the dark bottle losses to the change in the clear bottle, the gross photosynthetic production can be determined***

Measurement of productivity is a complex task about which whole texts have been written. However, it is an important aspect for the understanding of ecosystem function and management. Some key points will allow some assessment of significance to be placed upon information available in the literature. All primary production measurement is based upon the equation for photosynthesis:

$$6CO_2 + 6\,H_20 \rightarrow C_6H_{12}0_6\ \text{(glucose)} + 60_2$$

This equation shows that either the amount of carbon dioxide or water taken up, the amount of carbon converted into organic compounds or the amount of oxygen released could be used to measure primary production. Measurement of water is impracticable because of its general abundance but the other three are all potentially useful measures.

It is usual to estimate primary production (PP) in terrestrial systems by measuring the increase in plant biomass over a fixed time period but this only measures net primary production. This method is the 'harvest' method and is commonly used for crops and fields. Inherent problems include the difficulties of measuring root growth in perennial species and above-ground biomass of the woody parts of trees and shrubs.

In aquatic systems, measurement of gas exchange by the *light bottle/dark bottle* technique (Figure 7.3) has been used widely for measuring primary productivity. Samples of water containing phytoplankton are suspended at the experimental depth in clear glass bottles (light) and opaque, light-excluding dark bottles. Both photosynthetic production of oxygen and respiratory uptake of oxygen occur together in the light bottles (= NPP) while in the dark bottles only respiration is possible. By adding the oxygen lost in the dark bottle to the change in oxygen levels in the light bottle, a measure of gross primary production (GPP) results.

More sophisticated methods available today include spectrophotometric determination of chlorophyll abundance. This makes use of the observation that the rate of photosynthesis is proportional to the amount of chlorophyll present in the system and can be determined by either ground-based systems or remote-sensing methods. Under more experimental conditions, the radioisotope C^{14} can be incorporated into the carbon dioxide available for photosynthesis. This is then used to measure assimilated carbon by measuring how much of the radioisotope turns up in the carbon compounds of the plant.

Secondary production is rarely measured for whole communities because of its complexity but tends to be estimated from measures of net primary production. Figure 7.4 presents estimates for net secondary production calculated from net primary production data by making assumptions about the eating and digestive capabilities of the animals. It is therefore liable to considerable error. Energy flow through the detritivore– decomposer food web is even less well quantified and understood than secondary production within the grazing food chain/web system.

7.3.2 Comparative ecosystem productivity

Net primary production (NPP) of a system has now been estimated for most types of ecosystem (Table 2.1) and, although estimates do vary widely, a basic pattern of productivity emerges. Four main factors determine the productivity of terrestrial ecosystems, namely light, temperature, rainfall and nutrient levels. The high productivity of moist tropical regions is due to favourable combinations of high incident solar radiation, warm temperatures and abundant rainfall. Winter periods of low temperature and low light levels limit temperate and arctic regions while a shortage of water limits the primary production of the arid regions of the tropics. The really productive habitats are forests, marshes and swamps, estuaries and reefs, while deserts, tundras and the open ocean are relatively unproductive; most other systems, including agriculture fall somewhere between these extremes.

Photosynthetic efficiency, the ratio of NPP to visible light energy received, typically ranges between 1 and 2 per cent in terrestrial habitats. It is rarely as high in aquatic systems, where it is usually less than 0.5 per cent. Although photosynthetic efficiencies of 1 to 2 per cent occur in the most productive habitats, estimates suggest that plants assimilate only about 0.2–0.5 per cent of the light energy falling on the Earth. This is because most of the Earth's surface lacks the optimal conditions for plant growth (Figure 7.5). The total amount of energy received in a habitat varies with latitude and with prevailing weather conditions. Two-thirds of the annual primary

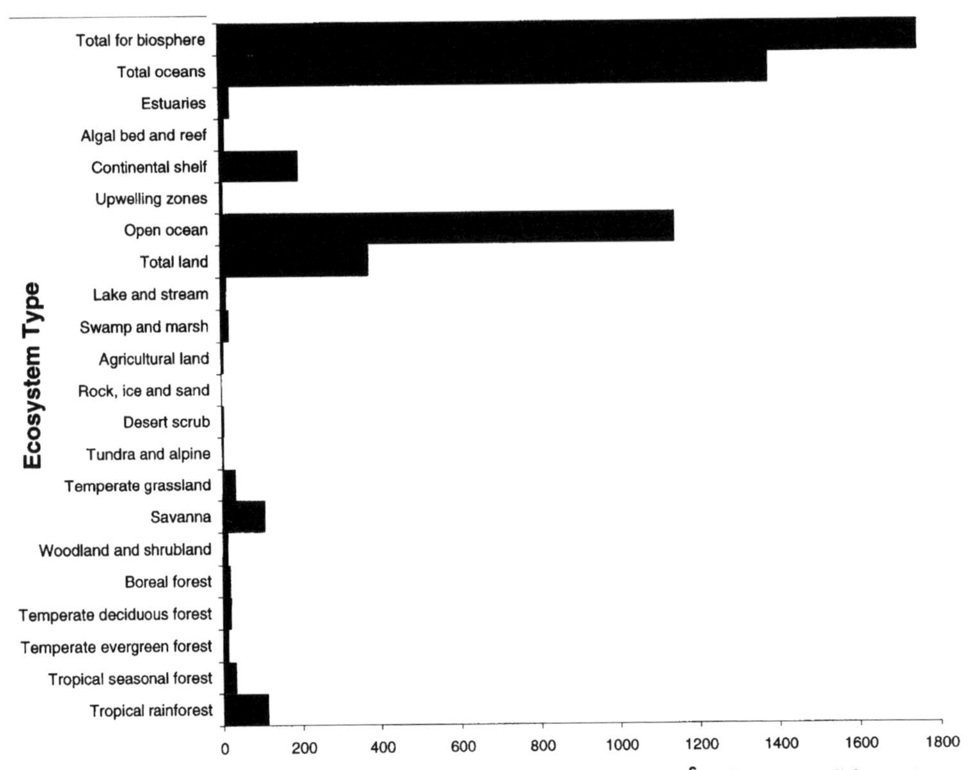

Figure 7.4 ***Variation of total net secondary production in the world's major types of ecosystem***

production of the Earth is terrestrial in nature and, although tropical forests cover only about 7 per cent of the Earth's surface, they account for about 25 per cent of total production. Temperate forests and the open ocean are responsible for between 14 and 25 per cent, respectively.

Secondary productivity is much less well quantified in all ecosystems and tends to be based upon calculations that themselves make somewhat gross assumptions about consumption and conversion of plant material into animal material. Perhaps the most notable feature, however, is the high consumption efficiencies assumed for marine systems. Marine herbivores are believed to be much more effective at grazing than terrestrial herbivores. The consequence is that the total secondary production of the oceans is almost the same as that for terrestrial habitats, despite the total primary productivity on land being about double that of the oceans. Estimates of oceanic secondary productivity are important for calculating the potential of fisheries.

Although most attention in productivity studies is paid to the animal and plant kingdoms and to the typical trophic interrelationships, more than 90 per cent of the energy from net primary production goes to the decomposer food web as detritus. On land, energy degradation by decomposers is far more important than that by all other organisms combined. Scavenging animals (detritivores) are common because so much of the primary production bypasses the herbivores and such organisms are vital to the material and energetic economy of most systems (Table 7.1).

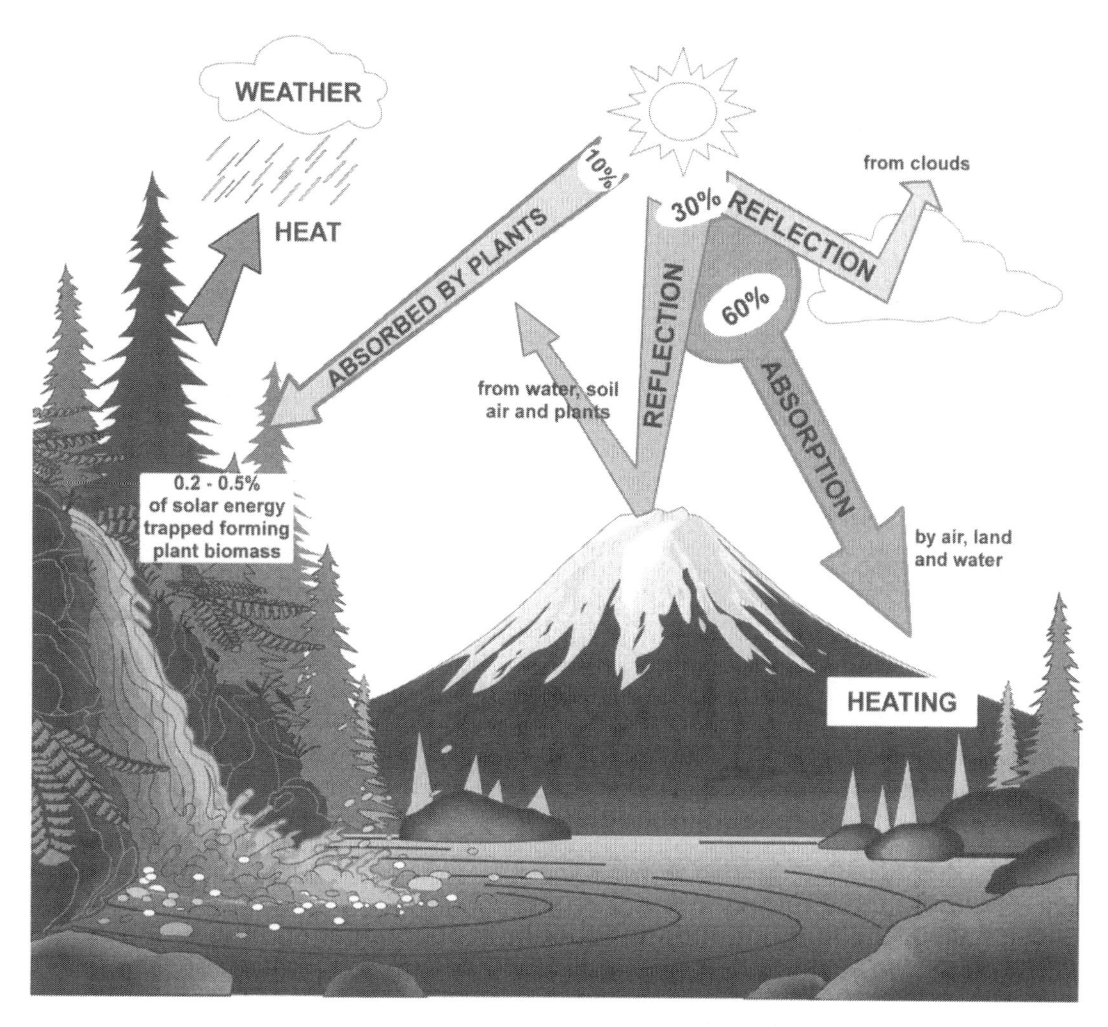

Figure 7.5 ***A summary diagram of the acquisition and loss of energy (sunlight) deriving from the sun***

7.3.3 Trophic interactions and pollution

One of the important consequences of trophic interactions is the transfer and recycling of materials between trophic levels. The transfer of materials along the food web is of considerable significance since it is by this means that many mineral, vitamins, etc. are obtained and conserved. However, many undesirable materials may also be incorporated into this process and this can result in the accumulation of materials to concentrations where a deleterious biological effect may occur.

Any chemical is potentially toxic if present in sufficient concentration. Elements most frequently associated with toxicity from environmental exposures include the heavy metals cadmium (Cd), mercury (Hg), chromium (Cr), silver (Ag), copper (Cu), cobalt (Co), iron (Fe), nickel (Ni), lead (Pb) and tin (Sn) and lighter elements such as aluminium (Al), arsenic (As) and selenium (Se). To these elements must be added the

Table 7.1 ***World detritus production, calculated from estimates of NPP***

Ecosystem	World NPP (dry wt) 10^9 tons/yr	Percentage of production going to detritus	World detritus (dry wt) 10^9 tons/yr	World carbon 10^9 tons/yr
Swamp/marsh	4.0	90	3.6	1.8
Tropical forest	40.0	95	38.0	19.0
Temperate forest	23.4	95	22.2	11.1
Boreal forest	9.6	97	9.3	4.6
Woodland/shrubland	4.2	80	3.4	1.7
Savannah	10.5	60	6.3	3.2
Temperate grassland	4.5	50	2.2	1.1
Tundra/alpine	1.1	95	1.0	0.5
Desert scrub	1.3	95	1.2	0.6
Extreme desert	0.07	97	0.1	0.03
Agricultural land	9.1	50	4.6	2.3
Totals	107.8		91.9	45.9

Source: Modified from Colinvaux 1993.

immense number of compounds, both natural and synthetic (organic chemicals mainly), known to exert toxic effects, e.g. DDT, malathion and tributyl tin.

Two trophically related processes exacerbate the problem of toxic materials in the environment, namely *bioconcentration* and *biomagnification* (see Section 3.2.4. and Box 3.7). Materials from the environment accumulate (concentrate) within an individual by a variety of biological processes usually associated with feeding or respiration. Thus the gills of a mussel act as an absorptive organ for heavy metals while they are being used for feeding and respiration. Such materials may then move to storage sites within the body where they may further accumulate and thus concentrate. Bioconcentration factors of several orders of magnitude occur for some substances, such as mercury.

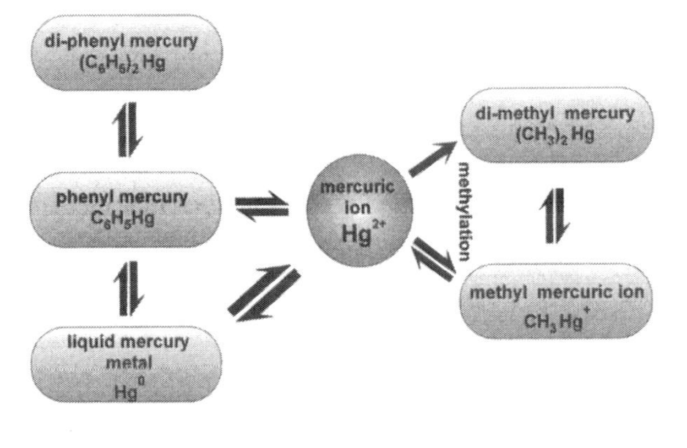

Figure 7.6 ***The biotransformations of mercury from metallic mercury (Hg^0) to the very toxic methyl mercuric form***

The second means of concentrating materials is through the eating of organisms, thus incorporating their body burden of the material into the body of the consumer. This process may occur at each trophic level so that, in the top predator, substances that seem innocuous at their original environmental concentrations may concentrate to levels at which toxic effects are significant (biomagnification). The story of DDT is a very good example (Box 3.7).

Another important biological process is the *biotransformation* of materials, often during the degradative metabolic processes of organisms, particularly micro-organisms. This is useful when it is part of the natural degradation and decomposition processes that are fundamental for materials recycling, a process vital to the sustainability of life. However, some transformations are deleterious. The transformation of relatively low-toxicity inorganic mercurial ions to the highly toxic methyl mercury ions (Figure 7.6) by microbial activities in the marine sediments of Minimata Bay in Japan is a good example (Box 7.2) but many more can be found in the literature.

7.4. Succession

Communities are always changing as organisms are born and die and energy and materials pass through them. This change in composition and form, primarily of the vegetation, is usually from simple to complex in a process known as *ecological succession*. This is *the non-seasonal, directional and continuous pattern of colonisation and extinction on a site by species populations*. It typically proceeds from immature, rapidly changing, unstable communities dominated by r-strategists to more mature, self-sustaining communities of K-strategists, providing that the process is not disrupted by major natural events or anthropogenic influences. The concept of succession began with the work of Clements (1916), who proposed that vegetational succession was a unidirectional series of changes that were irreversible. However, since that time, definitions of succession have been broadened and it is now recognised that they are sometimes reversible

A useful categorisation of types of succession is

- **autogenic succession**, a sequence of changes resulting from biological processes that modify conditions and resources;
- **allogenic succession**, occurring as a result of changing external geophysical forces;
- **degradative succession**, a third category of succession, also known as heterotrophic succession. This occurs over relatively short timescales and relates to the sequence of species changes occurring during the degradation of dead organic matter such as faeces, leaf litter or dead organisms. Degradative successions end because of the complete metabolism and mineralisation of the resource.

There are two types of autogenic succession, the concept of succession most widely discussed.

- **Primary succession** occurs on barren habitats where organisms gradually occupy an area and, in so doing, modify it. Such habitats are created by major geophysical events such as volcanic activity and glaciation and involve colonisation by *pioneer* species such as microbes, mosses and lichens that are usually r-strategists capable of rapid multiplication. Such species gradually modify the environment so that either new niches occur, a process called facilitation, or they inhibit further colonisation by other species, termed inhibition. Primary succession on land is *xerarch* succession while that which occurs in aquatic systems is *hydrarch* succession (Figure 7.7).

Box 7.2

The story of mercury: biotransformation, bioconcentration and biomagnification

Box 3.7 relates the story of the organic pesticide DDT, which was one of the first pollutants that drew attention to environmental processes leading to unanticipated environmental problems. The history of mercury pollution also reveals important processes related to the behaviour of metals such as lead, cadmium, copper and mercury. Mercury is widely used in a variety of industrial processes and products, the main ones including electrical apparatus; the chlor-alkali industry; production of paints and pigments; instrumentation and agriculture (as pesticides and fungicides). Many of these industries discharge significant quantities of mercury compounds to the environment. Mercury caused a number of physical and mental disorders in 1953 in Minimata Bay on the shores of Shiranui Sea, a part of the inland sea of Japan. Seventeen people died and twenty-three were permanently disabled and by 1975, 798 patients in various parts of Japan were diagnosed as victims of what became known as Minimata Disease.

The investigation of this incident revealed several critical features of the environmental cycling of mercury within that ecosystem:

- Although the principal releases were of the relatively less toxic inorganic forms of mercury (95 per cent of the discharge of mercury), the disease was due to accumulations of the most lethal form of mercury, the organic methyl mercury compounds. Methyl mercury is more dangerous because it is not excreted and acts as a cumulative poison causing progressive and irreversible brain damage. It was discovered that the inorganic forms were being converted to organic forms by microbial (methylating bacteria) transformation in the water and sediments (Figure 7.6).
- The methyl mercury then entered the food chain and was bioconcentrated at the various trophic levels.
- Food web transfers then produced biomagnification at each trophic stage so that, several steps up the food chain, the body burdens of mercury became sufficient to cause neuromuscular spasms and damage, kidney damage and birth defects. The magnification factors involved were of several orders of magnitude.

The recognition of these processes and their resultant problems resulted in forty areas in Sweden being closed to fishing, largely because of pollution from the chlor-alkali industry. In the USA, fishing was prohibited in various bodies of water in seventeen states. The processes of biotransformation, bioconcentration and biomagnification are now known to occur for a variety of other pollutant substances.

- **Secondary succession** is more common and begins with the disturbance, natural or anthropogenic, of an existing community. This includes habitats subject to man's impact such as burned and cut forests or abandoned farmland and which are important in debates about conservation and its methods or needs.

The exact sequence of species and community types that develop during succession can be highly variable. However, under ideal circumstances, the vegetation may pass through a series of stages that are characteristic of a given habitat and lead to a relatively stable form termed a *climax community*. The climax community of each region (see biomes, Chapter 2) depends mainly on climatic factors but is also strongly influenced by edaphic factors such as the frequency of fire and the grazing pressure

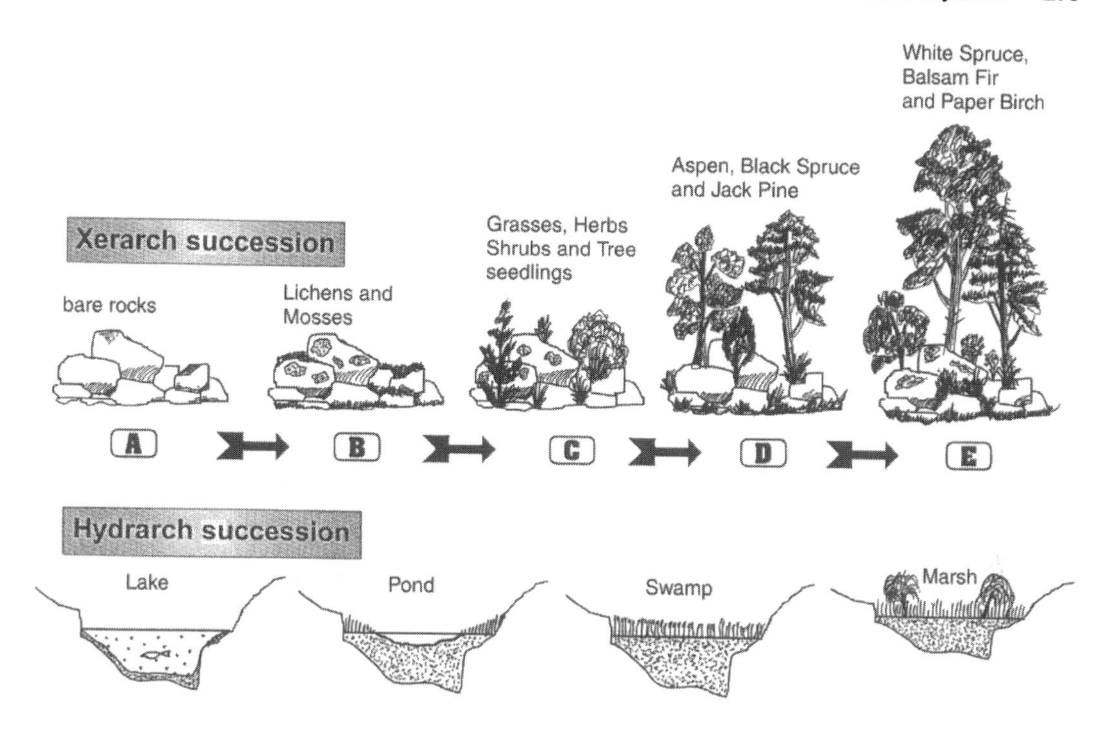

Figure 7.7 ***Diagrammatic representation of the main stages in the two main types of succession.***

(Table 7.2 and Figure 7.8). Succession is the inevitable result of plants with different life strategies living together in a region. Although it is described largely in terms of vegetation, succession obviously applies to the other components of the system. They are all interrelated and a change in the dominant vegetation will inevitably result in a change in the associated biota.

7.5 Biodiversity, stability and resilience

The Earth contains a rich and diverse array of organisms whose species diversity, genetic diversity and ecosystems are together called biodiversity. UNEP (1995) defines biodiversity as

> the variability among living organisms from all sources, including terrestrial, marine and other aquatic ecosystems and the ecological complexes of which they are a part; this includes diversity within species, between species and of ecosystems.

It comprises three distinct components:

- **Ecological diversity**: the diversity of ecological systems that do not really exist as discrete units but are part of a continuum. These include classifications determined by the scale being used and include biomes, bioregions, landscapes, ecosystems, habitats, populations and communities.

Table 7.2 ***Characteristics of immature and mature stages of ecological succession***

Property	*Immature system*	*Mature system*
Structure		
Species diversity	Low	High
Plant size	Small	Large
Trophic structure	Mainly producers, few decomposers	Producers, consumers and decomposers all represented well
Niches	Few only, mainly specialised	Many, mainly specialised
Community organisation	Simple	Complex
Function		
Feeding relationships	Food chains and webs mostly simple and mainly primary producers and herbivores with few decomposers	Complex chains and webs dominated by decomposers
Nutrient recycling efficiency	Low	High
Efficiency of energy usage	Low	High

Source: Modified from Miller 1995.

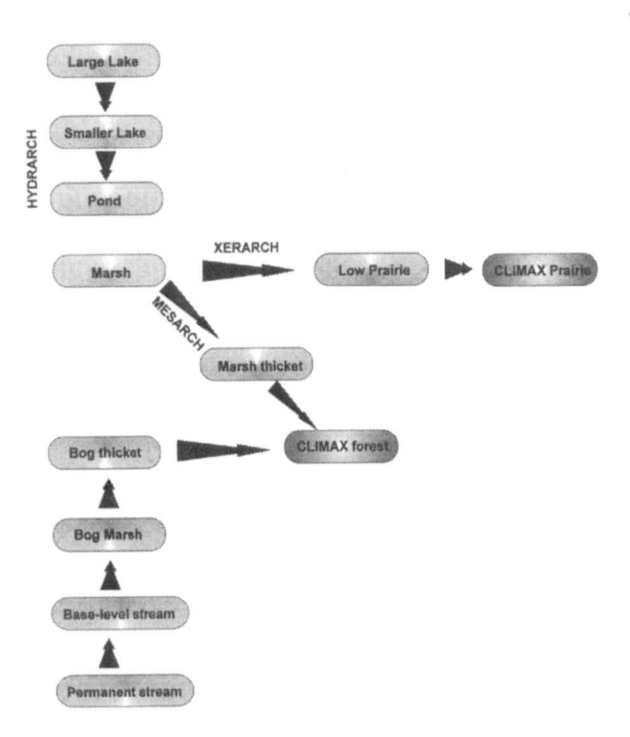

Figure 7.8 ***Examples of some basic successional sequences and their interrelationships***

- **Species diversity**: estimates of the total number of species on Earth average at between 13 and 14 million, of which only some 1.75 million have been described. Groups such as plants and vertebrates account for only 3 per cent of this estimated total, the majority belonging to groups such as insects, arachnids, fungi, nematodes and micro-organisms.
- **Genetic diversity**: variations in genotype provide the basis for all diversity. Variations in genotype are so great that species are genetically unique, yet there is genetic variability both within and between populations. Genetic diversity fluctuates in ways that are not completely understood but it does form the basis of adaptation and evolution of species and populations.

Many specialised, quantitative definitions of biological diversity have been developed as a means of making comparisons between communities. However, at its simplest level, it is a simple measure of the number of species found in the community, known as *species richness*. Most measures also include some estimate of the relative abundance or biomass of individuals amongst species, i.e. a measure of species dominance or rarity. This weighted measure of species richness is *species diversity*. Communities differ dramatically in these parameters since the number of species in a community varies with numerous factors acting on different temporal and spatial scales. The richest environments in terms of species richness appear to be:

- tropical rainforests, dominated by plant and insect diversity
- coral reefs where the diversity is spread more widely amongst animal phyla
- large tropical lakes where the high diversity is considered to be due to rapid evolutionary radiation in isolated but productive habitats.

Species diversity of most types of organisms increases towards the low latitudes, tropical forests containing over half of the world's species despite occupying only 7 per cent of the Earth's surface.

The number of species found in a community depends on:

- the variety and seasonal availability of resources;
- how specialised the species are;
- the amount of niche overlap.

More species may coexist if all available resources are usable and if niche overlap is extensive. The number of species also depends upon factors such as predator abundance, diseases, and the nature and frequency of habitat disturbances. In general, structural complexity of the community favours species richness when compared to less complex communities, e.g. grasslands. Disturbance also tends to enhance species richness, providing it is not too frequent or extensive, by making the habitat more patchy and increasing niche diversity.

Genetic diversity is the basic currency of evolution and adaptation and is affected by three significant processes today, all associated with human activities.

1. Habitat destruction, fragmentation and species/population extinctions can result in the loss of, or reduction in, genetic diversity.

2. Agricultural practices such as the development of monoculture of crops have resulted in displacement or extinction of wild varieties on a world-wide scale (Box 7.3).

3. Pollution from human activities has reduced both habitat and species diversity.

For most wild species, little information is available on the extent of the loss of genetically distinct populations and loss of genetic diversity. Reduction in population size will usually result in loss of genetic diversity and therefore adaptability of the species.

An important question for conservation and habitat management relates to the interactions between diversity and the concepts of *stability* and *resilience*. The stability of a system is an inherent property of its component populations and

Box 7.3

Monocultures in agriculture: efficiency with risks

Monoculture is a form of agriculture or forestry in which a single species, often selected for its productivity and/or disease resistance, is cultivated over a large area for economic efficiency. The economic justifications for such a strategy are fairly obvious but this process is fraught with risks. Most land is cultivated by replacing the naturally diverse communities with single crop or tree species. This often causes the leaking of nutrients from the land (resulting in the need to use chemical fertilisers) and, in regions such as the rainforests, the destruction of the soils. Much effort is also expended in protecting these crops from the inevitable invasions of pests. These include weeds, unwanted pioneer plant species which typically occupy the same niches as the crops; insect and other animal pests; and pathogenic (disease-producing) fungi, bacteria or viruses. The dangers of monocultures are clear when predator–prey interactions are considered.

Experience shows that simplified ecosystems are very vulnerable because they represent dense and abundant aggregations of the hosts of, often host-specific, parasitic or disease-producing organisms. The advantages of diversity and physical separation in diverse communities are lost. The result is a potential for massive population explosions of pest species that frequently have short generation times and rapid reproduction rates. The usual solution to this problem is the protection of crops, etc. by the use of pesticides or, more rarely, some form of biological control by a natural predator.

The use of diverse chemicals to overcome the problems of intensive methods of agriculture and forestry also poses its own problems, however. The substances used are frequently potentially toxic to a variety of other species, including man, and unless very judiciously used, represent a threat beyond the confines of their intended usage. Natural selection in rapidly breeding insect and other species of pests and disease-causing species has resulted in the development of genetic resistance to chemical pesticides. This leads to the use of stronger pesticide concentrations and the development of new toxic substances. The pesticides may eventually become ineffective, posing a severe threat to the host species.

communities and it is a measure of the ability of that system to accommodate environmental change. It is subject to the influence of a number of factors including size, response times and the stability of its component populations. Stability is conferred by the constant dynamic changes taking place within a community or ecosystem and, as such, varies between them but it is a very difficult parameter to measure. Three main components of stability are:

- **persistence** (inertia): the ability of a community or ecosystem to resistant disturbance or alteration
- **constancy**: the ability to maintain a certain size or maintain its numbers within limits
- **resilience**: the ability to return to normal following a disturbance.

It was believed that high species diversity conferred greater stability on a system since the number of dynamic links was greater and the system had more ways of responding to environmental stresses. However, research demonstrated numerous exceptions to this concept and it is now recognised to be a much more complex matter. At least a part of the problem lies in defining the concepts of stability and diversity. Thus Californian redwood forests and tropical rainforests have both high species diversity and high persistence, making them hard to disturb significantly. However, once degraded, their resilience is so low that the forest becomes hard to restore. Conversely, grasslands are usually of low diversity and low persistence but high resilience because their underground roots are difficult to destroy and allow rapid regeneration under most circumstances. Which of these is to be considered stable?

It is important to recognise that biological systems at all levels of complexity are rarely, if ever, at equilibrium. Natural systems, when subject to environmental variation, often change and come to operate within new criteria rather than return to some theoretical, optimal, equilibrium state. Disturbance and fluctuation dominate the real world so that the appearance of constancy and balance are largely superficial and are not the norm.

7.6 Biogeography

Biogeography is the study of the distribution of organisms, both past and present. It therefore includes the study of evolutionary histories as well as present-day distribution patterns. Some organisms found on most continents today may have been present together on Pangaea and subsequent continental drift has influenced their biogeographic patterns by effectively isolating them from populations of the same or related species. Evolutionary and ecological approaches characterise the two main subdisciplines of biogeography:

- **historical biogeography** investigates the evolutionary histories of taxa while
- **ecological biogeography** examines how ecological relationships influence the current distribution patterns of organisms.

These approaches must be integrated for a proper understanding of biogeography.

A species may occur in an area either because it evolved there or because it dispersed to the area. When a species population becomes isolated from other populations by the subsequent development of a barrier (e.g. by continental drift), the resulting distribution is a *vicariant distribution*. However, if they become separated because organisms have crossed an existing barrier, it is termed a *dispersal distribution*. This apparently clear distinction is often blurred by the fact that all organisms disperse and by the difficulty of defining barriers.

Taxa confined to a particular region are *endemic* to that location, and may be old taxa that are becoming more restricted in their distribution or new taxa recently evolved in a particular area. The effects of vicariant events are such that the longer the resulting isolation, the more endemic taxa it is likely to have produced. Australia is a good example where the isolation and evolution of the marsupials has led to large numbers of endemic species.

The Earth can be divided into six major terrestrial biogeographic zones (Figure 7.9) whose characters are maintained by dispersal barriers such as water, mountains and

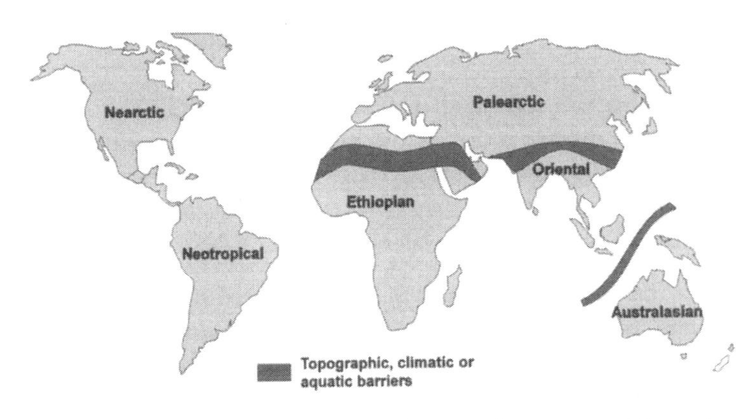

Figure 7.9 ***The six major terrestrial biogeographic zones and the major boundaries between them***

deserts. Within these are *habitat islands*, patches of habitat separated from each other by other habitats, e.g. alpine plant communities and deep-sea vent communities. There is, however, a convergence in appearance and ecological characteristics of vegetation living in similar climates in different parts of the world. These major types of ecosystems are termed biomes (see Chapter 2). Although characterised by their dominant vegetation and climatic conditions, it is important to appreciate that biomes also contain members of other kingdoms adapted to these particular biome conditions, i.e. they are integrated communities. The distribution of terrestrial biomes is strongly influenced by climate, especially temperature and rainfall.

Aquatic biomes, as outlined in Chapter 2, are subject to predator–prey, mutualistic and competitive interactions similar to those in terrestrial biomes. However, there are several differences which result from the properties of water and the restriction of primary production to the surface layers. Most aquatic photoautotrophs, the bacteria and algae, are microscopic in size and provide physical structuring within aquatic communities; only in shallow coastal waters do the macroscopic seaweeds provide a physically significant contribution. Animals, rather than plants, tend to be the dominant life forms in many aquatic communities, being the largest and most long-lived components. Indeed, in many of the deeper-water and less stable biomes, living plant material may be minimal or absent. In most aquatic biomes, water temperature, salinity and the nature of ocean currents replace the role of climatic factors in terrestrial communities. Fringing communities, however, may be subject to both climatic and aquatic parameters. Pelagic communities are dominated by planktonic and nektonic species and are relatively simple in structure while shore and shallow-water benthic communities are richer in species and more complex in structure. The richest and structurally most complex marine community is the coral reef that typically grows in warm but nutrient-poor waters: its diversity is founded upon the abundant niches which the variable coral structures provide.

Summary points

- A community is a dynamic, interactive collection of populations which occur together in space and time.
- An ecosystem comprises the biological community together with its physical environment and can be considered at a wide variety of scales.

- Communities are defined either by the habitat or by the dominant life-forms and also occupy a variety of scales from a rock-pool to a rainforest.
- Almost all populations and communities are distributed patchily because conditions and resources are themselves patchy. The dynamics of competition and predator–prey interactions vary spatially and temporally.
- Community interactions are founded upon trophic relationships and/or predator–prey relationships and may be very complex and subtle.
- Primary and secondary productivity varies markedly between ecosystems, terrestrial systems providing two-thirds of all primary production but only about half of secondary production. Decomposer systems are largely unquantified but are clearly very important.
- Changes in the composition and form of community vegetation follow distinctive patterns known as ecological successions, usually ending with a relatively stable climax community. Successions can be divided into primary and secondary types.
- Biodiversity includes habitat, species and genetic components and is under threat from human activities. It is difficult to measure but is an important parameter in any ecosystem.
- Stability of communities and ecosystems is a function of three properties: persistence, constancy and resilience.
- Biogeographic distributions are the result of both historical (evolutionary) and ecological factors. Terrestrial systems are dominated by climatic factors while aquatic ones have other determinants such as water temperature, current systems, and salinity.

Discussion / Further study

1. Select an ecosystem of your choice and list/describe the niches/microhabitats that are available within it. Include non-physical aspects of the niches, e.g. feeding relationships.
2. Assess the distribution of component species within a habitat of your choice. How will you sample this habitat and what significance does the distribution and size of each species have on the method of measurement?
3. Describe the succession you might expect to observe when (a) a new gravestone is placed into a graveyard, (b) a cultivated field is left fallow for several years and (c) a 10 square metre patch of rock on a rocky shore is cleared of all life forms and then left exposed to the environment. Explain your observations and conclusions in terms of the theory of succession.
4. Obtain from the literature estimates of the total number of species predicted to exist. What assumptions have been made and why do estimates vary markedly?
5. Now think about micro-organisms and their diversity. What is a bacterial species and how many might exist? Were they included in your answer to question 4?

Further reading

Global Biodiversity Assessment. V.H. Heywood (ed.). 1995. United Nations Environment Programme, Cambridge University Press, Cambridge.
A detailed and up-to-date overview of most aspects of global diversity assessment.

The Diversity of Life. E.O. Wilson. 1992. Penguin Books, London.
Superbly readable discussion of all aspects of diversity.

Environmental Issues: The global consequences. D. Mooney. 1994. Hodder and Stoughton, London.
A collection of simple but very clear expositions on the significance of ecosystem modifications related to man's activities.

Biodiversity: A biology of numbers and difference. K.J. Gaston, (ed.). 1996. Blackwell Science, Oxford.
Up-to-date critical evaluation of many aspects of biodiversity.

Environmental Ecology: The ecological effects of pollution, disturbance, and other stresses, 2nd edition. B. Freedmen. 1989. Academic Press, San Diego.
Very good examination of the topic with lots of case material.

References

Clements, F.E. 1916. *Plant Succession: Analysis of the development of vegetation*. Carnegie Institute of Washington Publication No. 242, Washington, DC.

Colinvaux, P. 1993. *Ecology 2*. John Wiley and Sons, New York.

Hutchinson, G.E. 1957. 'Concluding remarks', *Cold Spring Harbor Symposium on Quantative Biology* 22, 415–27.

Miller, G.T. 1995. *Environmental Science: Working with the Earth*. Wadsworth Publishing Co., Belmont, CA.

UNEP. 1995. *Global Biodiversity Assessment*, ed. V.H. Heywood. Cambridge University Press, Cambridge.

Glossary

accumulation *n.* The act or process of collecting together or becoming collected.

action spectrum *n.* A range of wavelengths of light within which a physiological process can take place.

alleles *n.* A particular form of gene. Alleles usually occur in pairs, one on each homologous chromosome in a diploid cell nucleus. When both alleles are the same the individual is described as being a homozygote; when each allele is different the individual is a heterozygote. The number of allelic forms of a gene can be many (multiple allelism), each form having a slightly different sequence of DNA bases but with the same overall structure. Each diploid form can carry only two alleles at one time.

allochthonous *adj.* Applied to rocks, detritus, etc., found in a place other than where they or their constituents were formed.

aposematic coloration *n.* Warning coloration in which conspicuous markings on an animal serve to discourage potential predators. Usually an aposematic animal is poisonous or unpalatable.

autecology *n.* The ecology of individual species (individuals and populations), including physiological ecology, animal behaviour and population dynamics.

autochthonous *adj.* Applied to material which originated in its present position, e.g. plant material such as peat, which actually grew where it was found rather than being brought in by outside influence, is said to be autochthonous.

bacteroid *n.* A modified bacterial cell, particularly of the type formed by species of *Rhizobium* within the root nodules of leguminous plants.

basal metabolic rate (BMR) *n.* The rate at which energy must be released metabolically in order to maintain an animal at rest. In animals, BMR is inversely proportional to body weight (i.e. small animals usually have a higher BMR than large ones).

capsid *n.* The outer protein coat of a mature virus.

chromosomes *n.* Microscopic rod-shaped structures that appear in a cell nucleus during cell division, consisting of nucleoprotein arranged into units (genes) that are responsible for the transmission of hereditary characteristics.

cladistics *n.* An approach to classification by which organisms are ordered and ranked entirely on a basis which reflects recent origin from a common ancestor (e.g. like a family tree). The system is concerned simply with the branching of the tree.

clone *n.* A group of genetically identical cells or individuals, derived from a common ancestor by asexual mitotic division.

community *n.* Any grouping of populations of different organisms that are found living together in a particular environment. The organisms interact and give the community a structure.

cross-fertilisation (or allogamy) *n.* The production of offspring by the female gamete being fused with a male gamete from a different organism.

crypsis (cryptic coloration) *n.* Coloration that makes animals difficult to distinguish against their background, reducing predation.

detritivore *n.* A heterotrophic animal that feeds on dead material (detritus). The dead material is most typically of plant origin, but may include the dead remains of small animals.

diapause *n.* A temporary cessation that occurs in the growth and development of an insect under the control of the endocrine system. Diapause is frequently associated with seasonal environments, the insect entering it during the adverse period and breaking from it when more favourable conditions return.

diploid *adj.* Applied to a cell nucleus containing two of each type of chromosome in homologous pairs and formed as a result of sexual reproduction. *n.* An organism in which the main life stage has cell nuclei with two of each type of chromosome, written as 2n. Diploid stages occur in all eukaryotes (except certain fungi) and allow a greater degree of genetic variability in individuals than the haploid state (n).

dominant (gene) *n.* In diploid organisms, a gene that produces the same phenotypic character when its alleles are present in a single dose (heterozygote) per nucleus as it does in a double dose (homozygote). A gene that is masked in the presence of its dominant allele in the heterozygote state is said to be recessive to that dominant.

dormant *n.* Alive but in a resting, torpid condition with suspended growth and reduced metabolism.

ecosystem *n.* A term used to describe a natural unit that consists of living and non-living parts, interacting to form a stable system.

ecotype *n.* A locally adapted population of a widespread species. Such populations show minor changes of morphology and/or physiology, which are related to habitat and are genetically induced. They can still reproduce with other ecotypes of the same species.

egestion *n.* The expulsion by an organism of waste products that have never been part of the cell constituents, i.e. non-digestible materials.

endoplasmic reticulum *n.* A complex network of cytoplasmic sacs and tubules which appears to be continuous with both the nuclear and cell membranes. If the membrane bears ribosomes, it is termed rough ER, and smooth ER when without ribosomes. Both are involved in the synthesis, transport and storage of cell products.

epibiota *n.* Plant or animal life growing on a surface.

epilimnion *n.* The upper layer of water in a stratified lake.

eukaryote *n.* An organism composed of cells that have a distinct nucleus enveloped by a double membrane and with specialised structures called organelles. All plants and animals are eukaryotes.

euryhaline *adj.* Applied to organisms that are able to tolerate a wide range of salinities.

evolution *n.* An explanation of the way in which present-day organisms have been produced, involving changes taking place in the genetic make-up of populations and being passed on to successive generations. Evolution is now generally accepted as the means which gives rise to new species.

facultative *adj.* Applied to organisms that are able to adopt an alternative mode of living, e.g. a facultative anaerobe is an aerobic organism that can grow under anaerobic conditions.

fertilisation *n.* The fusion of male and female gametes to give rise to a zygote which then develops into a new organism.

flocculation *n.* An aggregated woolly, cloud-like mass, usually of precipitated material.

gamete *n.* A specialised haploid cell (i.e. a sex cell) whose nucleus, and often cytoplasm, fuses with that of another gamete from the opposite sex or mating type in the process of fertilisation, forming a diploid zygote. In animals, male gametes are called sperm and female gametes eggs.

gametophyte *n.* A haploid phase of the life cycle of many plants, during which gametes are produced by mitosis. Gametophytes arise from a haploid spore produced by meiosis from a diploid sporophyte.

gene *n.* The fundamental physical unit of heredity that transmits information from one cell to another and hence one generation to another. A gene comprises a segment of DNA or RNA that codes for one or several related functions and that occupies a fixed position (locus) on a chromosome.

gene pool *n.* The total number of genes or the amount of genetic information possessed by all the reproductive members of a population of sexually reproducing organisms.

genet *n.* The term describing the persistent component of a modular organism which is of the same genotype, e.g. a colony of asexually produced organisms or a tree.

genetic recombination *n.* A rearrangement of genes during meiosis so that a gamete contains a haploid genotype with a new gene combination. Recombination is normally used to refer to rearrangement of linked genes on the same chromosome where the recombination is achieved by crossing over.

genetic variation *n.* A range of phenotypes for a particular character produced by alternative alleles of one or more genes, the range containing discrete groups or a continuous spectrum of types. Genetic variability arises initially by mutation and is maintained by sexual reproduction involving crossing-over in meiosis. This variation is the raw material for Natural Selection to act upon, ensuring that the best adapted variants are most likely to reproduce.

genotype *n.* The genetic constitution of an organism, as opposed to its physical appearance (phenotype). This usually refers to the specific allelic composition of a particular gene or set of genes in each cell of an organism, but it may also refer to the entire genome.

gizzard *n.* A muscular part of the alimentary canal where food is physically ground into small particles before digestion commences.

haploid *adj.* Applied to a cell nucleus containing only one of each type of chromosome. *n.* A haploid organism in which the main life stage has cell nuclei with one of each type of chromosome written as (n). Fungi and many algae usually have a brief diploid phase (2n) before returning to the haploid state by meiosis.

hemicelluloses or hexosan *n.* A polysaccharide associated with cellulose and lignin found in plant cell walls.

heterosis or **hybrid vigour** *n.* The increased vigour of growth, survival and fertility of hybrids as compared with homozygotes. It usually results from crosses between two genetically different, highly inbred lines. It is always associated with increased heterozygosity.

heterotroph *n.* An organism that is unable to manufacture its own food from simple chemical compounds and therefore consumes other organisms, living or dead, as its main or sole source of carbon.

heterozygous *adj.* (**heterozygote** *n.*) A diploid or polyploid individual that has alleles at at least one locus. Its phenotype is often identical to that of an individual that has one of these alleles in the homozygous state.

high-energy system *n.* A system where energy from cellular metabolism is used to maintain an elevated and constant body temperature.

holoplankton *n.* Zooplanktonic organisms that remain planktonic throughout their life cycle.

homeostasis *n.* The tendency of a biological system to resist change and to maintain itself in a state of stable equilibrium, e.g. the regulation of blood sugar levels by insulin.

homozygous *adj.* (**homozygote** *n.*) An individual that has the same genes at one or more loci. The phenotype of the genes for a particular locus will always be expressed since the two genes at the homologous loci are identical.

hypolimnion *n.* The lower, cooler, non-circulating water in a thermally stratified lake in summer. It is possible for the dissolved oxygen in the hypolimnion to deplete gradually since replenishment by photosynthesis and by contact with the atmosphere is prevented. Reoxygenation is possible only when the thermal stratification breaks down in autumn.

infauna *n.* Benthic organisms that dig into the sediment or construct tubes or burrows.

irritability (or sensitivity) *n.* The capacity of a cell, tissue or organism to respond to a stimulus, usually in such a way as to enhance its survival.

isomers *n.* Any two or more compounds that have the same molecular composition but different molecular structure. Isomers differ from each other in their physical and chemical properties.

lignin *n.* A complex, non-carbohydrate polymer found in cell walls, whose function is to provide stiffening to the cell, as in xylem vessels and bark fibres.

linkage *n.* In genetics, the association between genes located (linked) on the same chromosome, producing gametes with genotypes that do not occur in the ratios expected from independent assortment of the genes.

littoral *adj.* Pertaining to the shore.

longevity *n.* The period for which an organism survives.

meiosis *n.* A type of nuclear division (reduction division) associated with sexual reproduction. Four haploid cells are produced from a single diploid cell, involving two cycles of division. Meiosis has two major functions: (a) it halves the number of chromosomes to prevent a doubling in each generation, and (b) it produces a mixing of genetic material in the daughter cells by independent assortment and recombination.

meromictic *adj.* Applied to lakes in which part of the water column is stratified permanently, usually because of some physical or chemical difference (e.g. contrasting densities).

meroplankton *n.* Temporary plankton (i.e. the larval stages of other organisms).

mitosis *n.* A type of nuclear division by which two daughter cells are produced from one parent cell, with no change in chromosome number. Mitosis is associated with asexual growth and repair.

mutation *n.* A change in the structure or amount of the genetic material of an organism: usually applied to changes of individual genes but equally applicable to gross structural changes of chromosomes. Usually deleterious.

neritic zone *n.* The shallow-water or near-shore marine zone, extending from the low-tide level to a depth of about 200 m. This zone covers about 8 per cent of the total ocean floor and is the area most heavily populated by benthic organisms because of the penetration of sunlight to these shallow depths.

nucleotide *n.* A complex organic molecule forming the basic unit of nucleic acids, which are made up of three components a pentose sugar, an organic base (pyrimidine or purine), and a phosphate group.

obligate *adj.* Applied to an organism which has a fixed strategy with regard to a resource or condition. Opposite of facultative.

oceanic *adj.* Applied to the regions of the sea that lie beyond the continental shelf, with depths greater than 200 m.

osmoregulation *n.* The process whereby an organism maintains control over its internal osmotic pressure irrespective of variations in the environment. Such an organism is said to be an osmoregulator.

pectin *n.* A complex polysaccharide often found as calcium pectate in plants, where it is a component of the cell wall. When heated, pectins form a gel which 'sets', a feature used in making jams.

pelagic *adj.* **1** In marine ecology, applied to the organisms that inhabit open oceanic waters. **2.** In ornithology, applied to sea-birds that come to land only to breed, spending the major part of their lives at sea.

permafrost *n.* Permanently frozen ground on which only a thin surface layer thaws in the summer, as in the high Arctic and Antarctic.

phenetic classification *n.* The grouping of biological organisms on the basis of observed physical similarities.

phenotype *n.* The observable features of an individual organism that result from an interaction between the genotype and the environment in which development occurs. Organisms with the same overall genotype may have different phenotypes because of the effects of the environment and of gene interaction. Conversely, organisms may have the same phenotype but different genotypes as a result of incomplete dominance, penetrance or expressivity.

phylum *n.* A major grouping (taxon) into which kingdoms are divided and which is composed of a number of classes.

phytoplankton *n.* That part of plankton made up of plant life.

plankton *n.* The organisms that inhabit the surface layer of a sea or lake, drifting with the current and incapable of swimming from current system to current system. Members of the plankton vary in size from unicellular organisms to large jellyfish.

plastid *n.* A organelle found in plants and made up of a double membrane. Plastids are between 3 and 6 μm in diameter and have either a photosynthetic function (chloroplast) or a storage function (amyloblast).

polymictic *adj.* Applied to lakes (e.g. those in high altitudes in the tropics) whose waters circulate almost continuously.

population *n.* A group of organisms of the same species that occupies a particular area.

prions *n.* Specific infectious protein molecules that contain the information that codes for their own replication (e.g. BSE in cattle).

privation *n.* The loss or lack of the necessities of life, e.g. food and shelter, usually resulting from low densities of organisms, and the hardship resulting from this.

prokaryotic cells *n.* (pro = before; karyotic = nucleus) Simpler in design than eukaryotic cells, possessing neither a nucleus nor the organelles found in the cytoplasm of eukaryotic cells.

recessive allele *n.* An allele that only shows its effect in the phenotype when present in a homozygous condition. When paired with a dominant allele, the effect of the recessive allele is hidden.

recombination *n.* A rearrangement of genes during meiosis so that a gamete contains a haploid genotype with a new gene combination.

resistance *n.* Any inherited characteristic of an organism that lessens the effect of an adverse environmental factor such as a pathogen or parasite, a biocide (e.g. herbicide, pesticide) or a natural climatic extreme such as drought or high salinity.

salinity *n.* A measure of the total quantity of dissolved solids in water, expressed in parts per thousand by weight. Salinity is the degree of saltiness in a body of water and is 35 per cent in standard sea water.

self-fertilisation *n.* The fusion of male and female gametes from the same hermaphrodite individual. Self-fertilisation occurs, for example, in some parasites and some snails, but is most common in some plant groups.

sexually dimorphic *n.* Organisms whose sexes are visibly, externally different.

species *n.* (sing. and pl.) A group of organisms that resemble one another closely; the term is derived from the Latin *speculare*, 'to look'. In taxonomy, often used to distinguish groups of organisms that can interbreed within the group but which are reproductively isolated from other groups.

stereo-isomer *n.* One of two or more isomers that have the same molecular structure but differ in the spatial arrangement of the atoms in the molecule.

stratification *n.* (1) The layering of water in a water mass as a result of some factor which influences density. Often causes restricted circulation. (2) A process in which certain seed plants require a period of low temperatures, sometimes freezing, for germination to take place.

substrate *n.* (1) The reactant acted upon by an enzyme. (2) Any object or material upon which an organism grows or to which an organism is attached.

transpiration *n.* The loss of water vapour from a plant to the outside atmosphere which provides the plant with a circulation from root to shoot and which results in evaporative cooling.

upwelling *n.* An upward movement of water masses that results in nutrients being brought to the surface. Regions of upwelling are often regions of high ocean productivity, e.g. off the Peruvian coast, where there are major stocks of anchovies.

zooplankton *n.* The part of plankton made up of animal life.

Index

Printed in Great Britain by
Amazon.co.uk, Ltd.,
Marston Gate.